NUCLEAR LESSONS

RICHARD CURTIS AND ELIZABETH HOGAN
with Shel Horowitz

AN EXAMINATION OF NUCLEAR POWER'S SAFETY, ECONOMIC, AND POLITICAL RECORD

Stackpole Books

NUCLEAR LESSONS
Copyright © 1980 by
Richard Curtis
Elizabeth Hogan
Shel Horowitz

An Analysis of the Accident
at Three Mile Island
Copyright © 1980 by
Richard E. Webb

Published by
STACKPOLE BOOKS
Cameron and Kelker Streets
P.O. Box 1831
Harrisburg, Pa. 17105

Published simultaneously in Don Mills, Ontario, Canada
by Thomas Nelson & Sons, Ltd.

Parts of this book originally appeared as *Perils of the Peaceful Atom*
by Richard Curtis and Elizabeth Hogan (Doubleday, 1969).

Library of Congress Cataloging in Publication Data

Curtis, Richard.
  Nuclear lessons.

  Bibliography: p.
  Includes index.
  1. Atomic power-plants—United States—Safety
measures. 2. Atomic energy—Social aspects—
United States. I. Hogan, Elizabeth R., joint
author. II. Horowitz, Shel, joint author.
III. Title.
HD9698. U52C79   1979   333.79'24   79-25509
ISBN O-8117-1851-4

Printed in the U.S.A.

"Our technology has outpaced our understanding, our cleverness has grown faster than our wisdom."

DR. ROGER REVELLE
Chairman of the U.S. National Committee
for the International Biological Program

"Ours is a world of nuclear giants, and ethical infants."

GENERAL OMAR BRADLEY

"The men who create power make an indispensable contribution to the nation's greatness. But the men who question power make a contribution just as indispensable — for they determine whether we use power or power uses us."

PRESIDENT JOHN F. KENNEDY

Some say the world will end in fire . . .

ROBERT FROST

Contents

Preface

Ten years ago, when *Perils of the Peaceful Atom* was published, we issued the clarion call for shutting down the nuclear industry. Unfortunately, some of our predictions about the dangers of nuclear power have proved to be chillingly accurate. In the meantime, even more terrifying information has come to light about nuclear power. The facts accumulate steadily, and they continue to be no less than portentous.

The thesis of our original book still stands, its message more important now than ever before. For this new book we have tried to provide as much new information as possible in the hope that we will continue to fuel the fires of anti-nuclear protest and see our thesis become reality.

We are deeply indebted to many people for help and encouragement in writing this book.

Our lasting gratitude is extended to Dr. LaMont C. Cole, professor of ecology and systematics, Cornell University, who read the original text of the manuscript and contributed enormously helpful suggestions.

The original text also received invaluable guidance from Leo

Goodman, consultant for the United Auto Workers in Washington, D.C., who, having followed the development of nuclear power plants from their earliest days, has accumulated vast knowledge of the field, which he has shared most generously with many writers over the years —but never, we feel sure, more generously or helpfully than to the authors of this book.

Special thanks are also due to Larry Bogart, director of the Anti-Pollution League and co-founder of the Citizens' Committee for the Protection of the Environment, who has graciously shared his knowledge and experience with us, and who, like Mr. Goodman, has made untiring efforts to bring the hazards of nuclear power plants to public attention.

We wish to express special appreciation to Edward A. Friedman, assistant professor of physics of the Stevens Institute of Technology, for many valuable conversations in which he clarified certain technical points about nuclear technology.

We had substantial help in the updating of this book from Mary Jane Sullivan and Wendy Fisher, of New York Mobilization for Survival, Ernie Krumm of Great Falls, Montana, and Debbie Friedman of the Brooklyn Anti-Nuclear Group.

We are grateful, too, to Harold Grabau, Larry Grow, Ken Mc-Cormick, and David Curtis for their contributions to the original edition, to Scott Meredith and John Schaffner, the agents who originally arranged for publication by Doubleday, and to Martha Millard, the agent who arranged for publication of this updated edition.

Finally, our deepest appreciation to Stackpole Books, particularly to Neil McAleer and Andrea Chesman.

To the many other courageous individuals, too many to list, who have persisted in opposing nuclear power plants and publicizing their risks in the face of formidable obstacles and frequent frustrations, we are indebted not only as writers, but as private citizens.

Finally, it is impossible to express adequately our gratitude to Joanne Curtis and Margaret Hogan for their faith, patience, and many more tangible contributions to a book that would not have been possible without them.

RICHARD CURTIS
ELIZABETH HOGAN
SHEL HOROWITZ

Introduction
Lessons from Three Mile Island

March 28, 1979. In the early morning hours the fantasy that a major nuclear accident would never occur was shattered. At the brand-new Three Mile Island #2 nuclear plant, ten miles from Harrisburg, Pennsylvania's capital, a number of mistakes culminated in the most serious nuclear accident in the United States. The entire eastern seaboard, the most densely populated area in the country, became a hostage of fear. Never again could the nuclear industry's claims of absolute safety be believed.

Within days, enormous demonstrations were held around the world, demanding an end to nuclear proliferation. Demonstrators in Europe marched by the thousands under the slogan, "We All Live in Pennsylvania." The antinuclear battle had, after twenty-five years, become a mass movement. The world was frightened, and the American public had been betrayed.

Many questions remain unanswered, although a presidential commission has been appointed, and there is evidence of sincere soul-searching on the part of the Nuclear Regulatory Commission. The public and media, remembering the ten days of terror in early spring, will study the conclusions of these commissions with more than usual interest. Although the immediate crisis involved only Pennsylvania and surrounding states, almost everyone was aware that a similar accident (or worse) could occur anywhere near a nuclear reactor.

Among the major questions that come to mind now that the immediate crisis is over:

1. Could the TMI accident have been avoided?
2. Why did this accident occur?
3. What can be learned from Three Mile Island?

Taken in one sense the answer to the first question has to be yes — *if all the necessary precautions had been taken, and if conditions of nonaccident performance had been maintained.* This is, indeed, the foundation of the nuclear proponents' philosophy. References to "ideal" conditions are frequent in all testimony, literature, and public debate by the proponents. Because even they must recognize (however reluctantly) that ideal conditions are uncommon, they add the phrase "redundant engineered safeguards," which is supposed to compensate for any less than ideal conditions or circumstances. They are masters at "safety by skillful prose."

Dr. Hannes Alfven, a Nobel Prize winning physicist and a realist, has more accurately summed up the conditions necessary to make nuclear power safe:

> Fission energy is safe only if a number of critical devices work as they should, if a number of people in key positions follow all their instructions, if there is no sabotage, no hijacking of the transports, if no reactor fuel processing plant or repository anywhere in the world is situated in a region of riots or guerrilla activity, and no revolution or war—even a "conventional" one—takes place in these regions. The enormous quantities of extremely dangerous material must not get into the hands of ignorant people or desperados. No acts of God can be permitted."[1]

Although the Three Mile Island accident could have been avoided, others in the future (which also, theoretically, *could* be prevented) will *not* be prevented. And they will not because at one point or another they will violate the standards described by Dr. Alfven.

The answer to the second question—why this particular accident occurred and what actually happened—is clarified in part by the report in *Time*, April 9, 1979:

By the company's own version of events, there were at least five equipment breakdowns—of valves, pumps and fuel rods. Moreover, the engineers should have been able to cool the reactor to a safe level within twelve hours. Only after the reactor core has fully cooled and the abnormal radiation levels within the container building have been reduced will investigators be able to pinpoint the sequence of mechanical failure, as well as any human mistakes that might have been made. On that issue—human error—there are contradictory statements as well. . . .

But what is already evident is that the engineers at Three Mile Island, for whatever reason, were confronted by a situation they had not foreseen, were uncertain at first how to handle—and most unsettling of all, perhaps did not even understand.[2]

Just how little the experts understood the TMI accident was made clear by the release of transcripts of the NRC's closed-door deliberations of March 29 and 30. Roger Mattson, NRC director of reactor systems safety told the commissioners, "We have an accident that we have never been designed to accommodate."[3] For a while, faced with this situation, federal officials considered deliberate further destruction of the reactor, including starting a fire, to trigger another accident —one they believed they *could* handle.

One of the most disturbing elements of all was the lack of information available even to the experts to guide their crucial decisions. Governor Dick Thornburgh, for example, said of the problems facing him and NRC Chairman Joseph M. Hendrie: "We are operating almost totally in the blind, his information is ambiguous, mine is nonexistent —and—I don't know—it's like a couple of blind men staggering around making decisions."[4]

Senator Gary Hart (D-Colo), commenting on this, stated that the NRC was "totally unprepared for the kind of crisis which occurred. It was unable to verify the accuracy of information it was receiving from the site and to sort out conflicting information from the utility and state officials. All of this indicates a very serious gap in the NRC's capacity in crisis management situations."[5]

Other congressional representatives expressed concern over the apparent preoccupation of NRC members with unfavorable publicity. Representative Bob Carr (D-Mich), a member of two House nuclear energy subcommittees, is reported to have said, "They were far more interested in public relations than in public safety."[6]

It would appear that for the past several years the nuclear industry and utilities using nuclear power have suffered the same serious psychological blindspot. The following quotations from a booklet "Nuclear Energy: What is It All About?" prepared by Public Service Electric & Gas Company is unfortunately typical of many booklets that have been distributed over the years:

> It should be stressed that there is nothing in the design of nuclear power plants which involves significant technological unknowns. . . .
> Therefore we can design a nuclear reactor with engineering safeguards and use materials in its construction that virtually eliminate the possibility of accidental releases of radioactive materials. . . .
> Engineering safeguards in a nuclear plant assure complete protection against potential accidents. . . .
> Should the loss of coolant take place (an exceedingly remote possibility), the nuclear containment system is designed to prevent any fission products from being released to the environment.[7]

The irony of such assurances in view of Three Mile Island—combined with the statement that Harold Denton, director of operations for NRC, is reported on May 1 to have made to a Senate subcommittee that he still did not know how close the TMI plant came to a disastrous meltdown—can be matched only by subsequent efforts of nuclear proponents to minimize the March 28 accident.

The third and most important question is, What can be learned from the accident at Three Mile Island? In capsule form, we have learned that there were multiple possible causes of the accident:

1. *Design errors.*
   From an A.P. report, May 21:

> Congressional sources said a task force investigating the accident concluded that operators were unable to halt the nation's worst nuclear accident because their instruments conveyed data that was either wrong or hard to interpret.

2. *Inadequate training of operators.*
   John G. Kemeny, chairman of the Presidential Commission investigating TMI, quoted in the *Atlantic City Press,* June 1:

> We've heard again and again that the operators were faced with a situation that none of their training prepared them for. There seems to be overwhelming evidence of that.

3. *Violations by the plant's operator.*

U.P.I. report, August 3:

Stello, [NRC Director of Investigation Victor Stello] said he is considering 35 potential charges of violations against the plant's operator, Metropolitan Edison, for infractions of federal rules ranging from pre-accident leaks in the nuclear reactor to exposure of workers to high radiation levels.

4. *Failure of reactor manufacturer to respond to warning of unreliability.*

From an A.P. report, May 15:

Carl Michelson, a nuclear engineer for the Tennessee Valley Authority and an NRC consultant, wrote in January 1978 that the pressurized level is not considered a reliable guide as to core cooling conditions. . . .

Michelson said in a telephone interview from his home in Oak Ridge, Tenn., that his report was sent to Babcock & Wilcox on April 27, 1978, but he did not get a reply until early this year.

Two months later, on July 19, 1979, a special report from the *Washington Post* began:

Two nuclear safety engineers at Babcock & Wilcox Co. . . . said Wednesday they tried unsuccessfully to warn customers of the possibility of a nuclear accident similar to the one that crippled Three Mile Island. . . .

The engineers, Joseph J. Kelly and Bert M. Dunn, told the Kemeny Commission . . . that they wrote memos to Babcock supervisors and had repeated discussions with managers over what they felt was a need to warn Babcock customers of the possibility of a loss of coolant accident like the one suffered at Three Mile Island. . . .

The date of Kelly's first memo was Nov. 1, 1977, while Dunn wrote his first memo Feb. 9, 1978, more than a year before Three Mile Island.

5. *Failure of utility to correct a known defect adequately.*

From a U.P.I. report, May 31:

A critical valve that failed during the Three Mile Island accident malfunctioned almost exactly one year earlier during testing but was not adequately corrected, an official of the utility company that operated the plant testified Wednesday. . . .

In sworn testimony before the presidential commission investigating the accident Metropolitan Edison Company vice president John Herbein said after the March 29, 1978 failure of a pressure-relief valve, technicians

installed a light on the control panel to indicate that power had been trans-
mitted to close the valve. . . .

But the light did not indicate the valve was actually closed. "You went
only half way," said commissioner Carolyn Lewis.

6. *Human error—and fatigue.*

From a report by Larry Green, *Los Angeles Times,* April 5:

Simple human fatigue may have contributed to the lapse that led to the
most serious accident in U.S. commercial nuclear history. . . .

Maintenance technicians at the Three Mile Island nuclear power plant
worked 40 consecutive days without a day off in the period immediately
preceding the March 28 accident that officials now attribute in part to human error. . . .

And for all but the three days before the accident, maintenance crews
were working 10-hour shifts. . . .

In Washington, Wednesday, Nuclear Regulatory Commission (NRC)
officials confirmed earlier reports that workers had inadvertently left closed
crucial valves that should have been open to supply emergency cooling
water to the plant's two steam generators.

7. *NRC Failure to heed warning of reactor inspector.*

From a U.P.I. report, August 23:

A government inspector said Wednesday he warned his bosses again and
again about the danger of a mishap like the Three Mile Island nuclear accident, but all he got for his efforts was a bad performance record. . . .

James Creswell, a reactor inspector for the NRC, said his warnings went
unheeded because of "a certain mind-set, as it were, that these accidents
couldn't happen." . . .

Creswell told the President's Three Mile Island Commission he wrote
memo after memo questioning the adequacy of instructions for reactor
operators when nuclear plant pressure readings conflict. . . .

He said he issued the warnings because of the way operators at Toledo
Edison's Davis-Besse reactor performed in 1977 during two minor accidents
that resembled the TMI mishap.

It is particularly significant that Mr. Creswell attributes the failure of
NRC to take his warnings seriously because of "a certain mind-set . . .
that these accidents couldn't happen."

The same point was made by two Nuclear Regulatory commissioners, Peter Bradford and John Ahearne, in an interview in *Critical
Mass Journal,* July 1979:

CMJ: Would it be fair to say that both the Commission and the utility were not prepared for an accident such as TMI?

Bradford: I think so. . . . It's as John has said on other occasions, one slips into a mentality in which accidents won't happen.

A little later in the interview, Commissioner Ahearne asserted:

That goes back to the point that Peter referred to earlier—my theory that *the whole system was just set to believe that accidents wouldn't happen. When you believe that, then there are things that you don't do.*

The tendency to take even the most serious warnings casually— "not to do things"—goes back a very long way, to the NRC's predecessor, the Atomic Energy Commission. A Jack Anderson column of June 21, 1979, revealed, for example, that the AEC had failed to follow through on a warning by its own safety experts in a report dated September 5, 1969. This report stated:

Hydrogen gas would be produced as a consequence of a loss-of-coolant accident. We are currently reviewing the problem of hydrogen production and several methods for control of the hydrogen concentration for all reactors, and have not yet established the methods which will be acceptable.

As Anderson commented, after giving this quotation:

Having posed the problem—and noted that they didn't have an answer to it—the safety officials incredibly decided it was okay to let things slide.

In the same column Anderson made other especially telling points, referring to TMI accident:

Even the plant's emergency system, designed to render excess hydrogen harmless, was not hooked up, the NRC's Dr. Roger Mattson told our associate Howard Rosenberg. The government had to fly planeloads of lead bricks in as shields before the emergency system—two hydrogen recombiners—could be made operational. And one of the two recombiners subsequently broke down. . . .

Incredible as it may seem, the NRC's Advisory Committee on Reactor Safeguards assured Congress in January 1978 that hydrogen control was one of a number of inherent problems that had been "resolved." . . .

But in the peculiar jargon of bureaucracy, "resolved" is in no way the same as "solved." As a memo accompanying the NRC report explained:

"In some cases an item has been resolved in an administrative sense." In other words, *the problem had been resolved only on paper, not at the reactors, where it counts.*

This confirms the point about the nuclear industry depending on "safety by skillful prose."

In the months since the accident, other defects in regulation and procedure that directly affect public health and safety have been revealed. These include *the NRC's failure to insist on adequate evacuation plans,* a failure contained in a congressional report critical of NRC's role in planning for reactor accidents,[8] and *the utility company's failure to inform citizens in advance of deliberate radiation releases.*

A further and serious weakness of the present "system" of public protection was summed up in a *Washington Post* report of May 20, 1979, by Representative William F. Goodling (R-Pa): "Most of my constituents cannot understand how their children could have been permitted to be exposed during those early hours while waiting for school buses or while on the playground. They believe it is sheer negligence to have permitted this to happen when an earlier notification would have kept them indoors." It is interesting to note that Congressman Goodling also told the President's commission investigating the accident: "I have talked to 1,500 of my constituents at five public meetings and I don't believe people realize what psychological effect this accident has had on them. They do not want the damaged core permanently stored on Three Mile Island. In fact, many are adamantly opposed to the reopening of Three Mile Island as a nuclear generating plant."

Sadly, but not too surprisingly, the desires of the people who are still endangered by the plant are being steadfastly ignored. In fact, regular releases of supposedly "safe" amounts of radioactivity are being emitted into the air as part of the rehabilitation of the crippled reactor (although it is conceded that no one knows exactly how much was originally released because the monitoring was inadequate). Meantime, the owners of the TMI nuclear plant have declared that without at least a $33-million rate increase they will face bankruptcy. While it is not absolutely certain, it appears likely that the very customers most endangered by this power plant (including those who are firmly opposed to its "rehabilitation") will be stuck with multimillion-dollar repair bills.

Of greatest concern, however, is the fact that most other utility companies are probably no better prepared to handle a major nuclear accident than the operators of TMI were. Referring to Metropolitan Edison Company, NRC Commissioner John Ahearne stated in an April interview: "It's not yet clear to me that Met Ed is unusual in this situation. An operating utility doesn't have the backup staff to look at the questions in an accident this severe."[9]

Candid assessments of various other nuclear power plants by the NRC's own inspectors, uncovered by the Union of Concerned Scientists through the Freedom of Information Act, supply further food for sober thought:

On Zion (Illinois):
Safety is substantially worse because of poor attitude and marginal management. Inadequate management controls. Management lacks ability to discipline employees for operator errors and carelessness. . . . Stability of staff a problem. Attitude regarding safety is poor.

On Robinson (South Carolina):
Licensee reports only those items that are conspicuously reportable. Licensee impedes inspector access and freedom of movement at site. No information is freely given. Does only what is required.

On Peach Bottom (Pennsylvania):
This is the least safe site in Region I and has the poorest management. QA (Quality Assurance) and security are not upgraded to current standards. Many repeat items of non-compliance (with federal regulations). Plant staff has appeared incapable of correcting increased radiation levels. Management is slow responding to problems. . . . Careless operations and poor maintenance.

On Oyster Creek (New Jersey):
They tend to just meet the minimum requirements. Design review of this plant was deficient. Plant was built at minimum cost. Rad waste, fire protection, and system separation are inadequate.[10]

Additional cause for concern are points made by the Union of Concerned Scientists in the May 1979 issue of *Nucleus:*

1. TMI-2 came perilously close to a meltdown—Had the reactor been in operation at full power for a year rather than several months . . . a massive core meltdown would have occurred during the early hours of the accident. It was only because the inventory of hot fission products was

not very great that the core did not heat up much more rapidly. Even so, as many as 50% of the fuel rods in the core may have suffered some damage.

2. The emergency systems were unable to deal with the accident. . . .

3. The plant was saved by non-safety related equipment. The systems actually used to cool the reactor were those very systems assumed to *fail* in the event of a serious accident. Had the main reactor coolant pumps broken down during the course of the accident, for example, all ability to cool the reactor would have been lost.

4. The plant was not designed to handle the series of events which occurred at TMI. . . . Ultimately, the plant operators were forced to improvise in order to prevent a core meltdown.[11]

Despite the evidence already assembled, nuclear apologists have mounted an expensive public relations program to combat concern over the TMI accident. This program is characterized by two arguments: (1) no one was killed, and (2) the system was designed to control accidents — and this is what it did. *Critical Mass Journal* raised the question of whether this is an intelligent response to NRC commissioners Bradford and Ahearne. *Their* responses are worth keeping in mind too:

> *Bradford:* I guess I would hate to think that the final, ultimate conclusion to come out of Three Mile Island would be that all the safety systems worked and therefore there's nothing to worry about.
>
> *Ahearne:* I think that one of the major conclusions is that each system *didn't* work.
>
> *Bradford:* Without attributing that particular line of reasoning to a given company or a set of companies, I do wince when I see both of those statements advanced as being the lessons of Three Mile Island.

Before Three Mile Island forced the dangers of nuclear plants into the public eye, the nuclear industry made repeated claims of safety. Carl Walske of the Atomic Industrial Forum, for example, bragged that the nuclear industry's safety record has been "impeccable."[12] Our own examination of the record shows that atomic proponents are either in desperate need of stronger glasses or are simply continuing the public relations campaign to promote nuclear industrialization — and the facts be damned. Deception, misstatements, and outright lies are not new weapons in the utility industry's fight to saddle the public with nuclear power.

In fact, literally hundreds of incidents, abnormalities, and anomalies occur every year. The accident record had become thick until a provision in the Energy Reorganization Act of 1974 changed the definition of a reportable accident to cover only those incidents where potential radiation release or catastrophic accident could be proven. As a consequence of this semantic twist, the first six months of 1975 had only 6 incidents recorded, compared with 850 during 1973.[13] The 1973 list showed as many as 65 incidents at the Browns Ferry reactor in Alabama, which came within inches of a meltdown two years later. The 850 figure, averaged over thirty reactors operating that year, yields a mean of 28 abnormalities per reactor—more than one incident per plant every two weeks. Many accidents would not be given a second glance in a more traditional industrial facility. Small component failures, cut fingers, and even doors open at inopportune times (such as at the Vermont Yankee plant, where the reactor was once inadvertently started with its lid off and the containment vessel door ajar)[14] must all be taken into account.

Much of this is the result of sheer negligence on the part of utilities, which are not penalized when they maximize profit at the expense of safety.

> During the year ending 30 June 1974, the AEC found a total of 3,333 safety violations at the 1,288 nuclear facilities [including all stages of the fuel cycle] it inspected. Ninety-eight of these posed a threat of radiation exposure to the public or to workers. The AEC imposed punishments for only eight of these violations.[15]

Nor is this purely a domestic problem. Accidents have happened in Canada, England, Japan, and in other countries. Furthermore, reactors in some parts of the world are being built without even the minimal attention given to safety. In the Soviet Union, for example, power plants are being constructed with *neither an emergency core-cooling system nor a steel-reinforced concrete containment vessel*—perhaps the two most important safety devices in American reactors.[16] In other words, the Soviets have made no provision for preventing or containing a core meltdown. (Not that our own precautions should be taken very seriously; when the government finally *tested* a scale model of the Emergency Core Cooling System, it failed six out of six times.[17])

A brief look at a few of the most dangerous accidents so far is illuminating:

• **Three Mile Island, Pennsylvania, March 28, 1979:** Pump failure caused large radioactivity releases. Young children and pregnant women were evacuated; the surrounding population was almost evacuated.

• **Browns Ferry, Alabama, March 22, 1975:** Use of a candle to check for air leaks caused a serious accident with a potential meltdown in two large reactors.

• **Traverse Bay, Michigan, January 1971:** A B-52 bomber flying over a Michigan reactor crashed a mere two miles from the site. The consequences if the plane had been carrying nuclear bombs, or if it had crashed directly into the reactor, are not pleasant to contemplate.

• **Hanford, Washington, September 30, 1970:** A short circuit caused a new electrical circuit to form, cutting the reactor off totally from its *scram* (immediate shutdown) mechanism. Disaster was averted because the reactor was equipped with a backup scram system of a different design, which did lower the eighty-seven control rods into the reactor's guts. This was a government reactor; commercial reactors are equipped with only primary scram systems and do not have a secondary method. In at least one case a boiling-water reactor's scram system was found completely inoperative in a routine test.

• **Lucens, Switzerland, January 21, 1969:** Large amounts of radioactivity were released after a loss-of-coolant accident. The reactor was located in a cavern, which had to be sealed off.

• **Enrico Fermi, Michigan, October 5, 1966:** The first commercial breeder suffered a partial meltdown when a last-minute safety addition worked loose and blocked the coolant flow. A nuclear explosion could have resulted.

• **Shippingport, Pennsylvania, 1964:** Replacement steam generators, heavier than the original, were installed with faulty supports. They fell as they were being filled with coolant, preparatory to reactivation of the reactor. Had this happened when the reactor was operating, a severe loss-of-coolant accident would have occurred.

• **Idaho Falls, Idaho, January 3, 1961:** Three men were killed during a nuclear runaway at a small test reactor.

• **Windscale #1, England, October 7, 1957:** A fire almost caused an explosion. Radioactive milk from a 200-mile radius was dumped into the Irish Sea.

• **Idaho Falls, Idaho, November 1955:** A core meltdown at small experimental breeder reactor. The reactor came within a half second of exploding.

• **Chalk River, Canada, December 2, 1952:** A partial fuel meltdown released more than a million gallons of radioactive water.[18]

There have been many, many others. Often, the only reason why an accident did not have truly catastrophic consequences was the combination of low capacity and remote location. We are now faced with a far more deadly situation: Giant reactors are going up just outside our major cities, often in densely populated suburbs.

Surely, we are being far more foolish than Prometheus.

1

The Goose That Laid
the Radioactive Egg

The Three Mile Island reactor, designed by Babcock & Wilcox, is a Pressurized Water Reactor (PWR). In a PWR, water circulating through it is kept under extremely high pressure to prevent steam from forming. Without pumps working continuously, the pressure cannot be maintained. The PWR and the Boiling Water Reactor (BWR) are the two common types of nuclear reactors built in the United States. The BWR is not immune to problems, and, in fact, it stands a better chance of allowing radioactivity to escape through the cooling water, for it has one less loop to contain the water, thus creating a higher chance of seepage into the environment.

A third type of reactor is the liquid-metal fast breeder reactor. The fast breeder has several essential differences from the more common PWRs and BWRs. While most reactors are cooled with water, the fast breeder uses volatile liquid sodium (which can explode on contact with air or water). Furthermore, the breeder is expected to generate, or breed, more fuel than it consumes by changing an inert isotope of uranium into plutonium fuel. Another difference is in the concentration of fuel. Enough fissionable material is concentrated in the breeder to allow the unnerving possibility of the formation of a "critical mass," leading to a small atomic explosion, an event that would be impossible in water-cooled models.

The Enrico Fermi Power plant was the first attempt at a commercial breeder reactor. Its history has been somewhat similar to Three Mile Island's: there were many unheeded warnings that accidents could happen. Until finally, one major accident did occur. It seems from the Fermi experience that too little was learned too late.

Lagoona Beach, Michigan, is located on the western shore of Lake Erie, near the city of Monroe, population 20,000. The surrounding farmland, irrigated by the Raisin River, is rich, the scenery agreeable, and the beach resorts attractive. It's no more than an hour by car to downtown Detroit; a radioactive cloud takes a little longer, unless there's a strong wind.

For thirty days in the autumn of 1966, it would have been difficult to find a reliable nuclear physicist prepared to wager that a radioactive cloud was an impossibility. For something immeasurably frightening had happened at the Enrico Fermi Power Plant on October 5,[1] and scientists, engineers, utility executives, and members of the Atomic Energy Commission held their breath while viewing a situation that raised the possibility that Detroit with its more than 1.5 million inhabitants might have to be evacuated.

On some eight hundred acres in Lagoona Beach the plant stood, dedicated to the memory of the Italian physicist whose prodigious intellect had been largely responsible for the first controlled release of atomic energy. By invoking Fermi's name the plant's promoters had undoubtedly hoped his spirit would smile on their project, which promised consequences as significant to world peace as those the atomic bomb had had to world war. Yet the history of the enterprise seemed to suggest that the late Nobel Prize winner would have difficulty complying with their petitions.

At the heart of Fermi was an atomic reactor, fashioned to generate enormous heat through controlled fission of the uranium fuel in its core. The heat would convert water into steam, which would in turn operate electricity-producing turbines.

In many ways it was an experiment. But one might have rested easier had earlier test reactors employing these principles and features proved reliable. Furthermore, trial runs at Fermi had given no one reason to hope that this machine would give a better accounting of itself than any other. The plant had been plagued by mishaps since completion, the latest of which revolved around the system by which molten sodium is circulated through the tubes containing uranium fuel, cooling them and carrying off heat to the boilers. Liquid sodium is at best a tricky substance, and at worst a violent one, and it had been misbehaving in Fermi's core. Finally, complications in the steam generator, where the sodium system heats the water, had caused a prolonged shutdown.

By the fall of 1966, however, the apparatus seemed to be back in working order, and on the evening of October 4 clearance was given to start up the reactor. With the utmost delicacy, technicians in the control room activated machinery for withdrawing graphite rods from the reactor's core. These rods absorb radioactivity; when withdrawn, they increase the fuel's rate of chain reaction and thus its heat. Withdrawal of the rods raised the temperature of the reactor to a little beyond 550 degrees F., at which point the engineers leveled off operation for the night.

On the following morning, after discovery and repair of a faulty valve, they set about boosting the power. Slowly, slowly, withdrawal of the rods resumed. Around two in the afternoon a pump malfunctioned, but by 3 P.M. a replacement was activated and they brought the power up to 20,000 kilowatts of heat energy.

At this time one of the instruments in the control room registered erratic fluctuations. Though not necessarily ominous, they led the technician to conclude it might be a good idea to switch from automatic control of the rods to manual for better handling if trouble should develop. When the signal steadied off, he went back to automatic, pushing the power up to 34,000 kilowatts, or 17 percent of maximum capacity. The fluctuations reappeared on his instrument. In addition, several temperature recorders showed excessive heat in certain spots in the core. And an observer noticed that the control rods appeared to be withdrawn farther than they should be for the desired heat level.

While they were wondering what to do next, the bells went off.

Both inside the reactor and in some adjoining buildings, sensors were picking up high radiation readings, setting off alarms throughout the complex. Automatic devices sealed off every area where high radiation was manifest. Luckily, no one was in those areas. But there were people outside—about two million of them within a radius in which fatalities could conceivably occur. And some radioactive gas, it was later determined, was released during the accident.

An agonizing length of time seemed to pass before the word went out to "scram" the reactor. Accordingly, six safety rods were thrust into its bowels to absorb radiation and kill a potential runaway chain reaction. Then everybody breathed slowly, waiting for the thing to go one way or the other.

Undoubtedly, in the agonizing moments that followed, many present at Fermi wondered about matters they had never really wondered about until that moment, particularly what it was like to die from radiation poisoning. Some may have been thinking of those poor souls at SL-I, the experimental reactor in Idaho Falls, Idaho.[2] In 1961 the plug had blown off its fuel core, sending three men to a particularly gruesome death. The first was dragged out by a heroic rescue team who suffered a three or four years' dose of radiation from the few moments in which they were in the contaminated chamber and in contact with the victim, who died a few minutes later, apparently of a head injury. But the building had been filled with so much radioactive poison it took more than five days before rescuers could recover the last of the bodies. If the explosion itself didn't kill the fellows, they must have been roasted alive by fission products. It was reported that highly radioactive portions of their bodies were removed and buried in the hot-waste dump at the site; the rest were put in small lead boxes and placed in caskets for burial.

But quick death was a blessing compared to the hideous torture of slow radiation poisoning. It would start with nausea and vomiting; the bowels would turn to water, then to blood. Fever, prostration, a gamut of mental symptoms ranging from stupor to hysteria. Then the burns would appear. Reduction of white blood cells would cancel bodily defense against infection, initiating internal and external bleeding. The hair would come off in obscene patches. Delirium, then death. If not death, something worse: life. Permanent emaciation, crippling, recurring symptoms, unthinkable pain, vulnerability to further disease and infection, cancer or leukemia, shortening of life expectancy, sterility or damage to reproductive organs leading to genetic defects in offspring.

Others at Fermi, perhaps, were able to turn their thoughts away from themselves and reflect on broader consequences. *Would* they have to evacuate Detroit? What plans had been drawn up by the city and state governments to effect the orderly withdrawal of up to two million people from their homes? Where would they go? Who would remain behind to protect their property? What would happen to their business? What would happen to *the* business of Detroit, the automobile industry? How long would they have to remain in exile? What radiological procedures and facilities existed to handle victims? The questions begot more questions until the prospects took on the proportions of a stupendous nightmare. Their only contact with sane reality lay in the radiation gauges and badges in their control room: So far, these did not register contamination.

A cautious investigation indicated that a runaway had been averted by scramming, the emergency insertion of the control rods. But the crisis was far from over. On the contrary, the initial diagnosis was that some of the reactor fuel had melted, a matter of profound gravity. When a critical mass of uranium 235 is collected in one place, it undergoes a violently spontaneous chain reaction. At a hasty meeting the fear was expressed that enough uranium had recongealed so that a disturbance of the core—by an attempt to remove the damaged fuel, for example—would jar it into a critical mass too great to be controlled by the control rods, which were already at their maximum. The stuff could explode, not with the force of an A-bomb, but still with sufficient impact to breach the steel and concrete containment structures of the building and hurl clouds of lethal gas into the sky. Then, depending on which way the wind blew, and how strongly, and how much material escaped . . .

Walter J. McCarthy, Jr., assistant general manager for the company that had developed Fermi, attended that meeting. He was later to say that the possibility of such a secondary accident was "a terrifying thought."

For the next month technicians and experts tiptoed around Fermi and spoke of its ailment in whispers, like aborigines camped on the slopes of a volcano, fearful of provoking the earth god's wrath. And though they were to proceed with excruciating deliberateness, they had no way of knowing whether their next probe might not bring the god raging to his feet.

In one sense Fermi was a special reactor. Its potential for "breeding" more fuel than it consumed was seen as answering many of the world's fuel-energy problems for centuries to come. Indeed, because the entire future of atomically produced electricity depended on the success of breeders — if they failed, low-cost uranium supplies would be exhausted before the end of this century — Fermi represented some of the highest expectations of government, industry, and tomorrow's electricity consumer.

Despite the Fermi accident, the dream of creating fuel where none existed before, the dream of a self-generating power source, is far from dead. President Carter and Congress have been struggling since 1977 over plans for a breeder at Clinch River, Tennessee; the President has resisted the breeder not out of fears of accidents, but out of concern over nuclear weapons proliferation. At present the only operating full-scale breeder plant is the Phenix, a 250-megawatt (Mw) facility in Pierrelatte, France, although a 1200-Mw Superphenix is under construction at Creys-Malville.[3]

It had been expected that the *doubling time,* the time it takes the initial load of fuel to double would be as low as just a few years. In actual operation, however, the Phenix doubling time will be between forty and sixty years,[4] far too late to be able to refuel current reactors. Nor is a workable commercial breeder program expected in the United States before at least the 1990s.[5]

In another sense, though, Fermi shared almost all of the problems and treacherous hazards confronting every other reactor designed before and since. The Fermi story is especially representative, however, because almost all of the issues raised in the pages that follow were raised, futilely, in a battle over that power plant that proceeded to the highest court in the land — where they were shockingly resolved in favor of atomic power's proponents. That the menaces of which the public was then warned are today growing to monstrous proportions makes it imperative that the lessons inherent in the Fermi episode not be lost.

Early in the 1950s, some thirty-five utilities and equipment manufacturers formed a company to design the Fermi plant,[6] and the following year these firms were assembled into the Power Reactor Development Company (PRDC) for the purpose of constructing, owning, and operating it. Late in 1955 PRDC filed its application for a construction permit with the AEC and humbly lowered its head in anticipation of applause from a public grateful to be getting atomically produced, low-cost electricity in the near future, and delighted at having a pace-setting scientific facility for a neighbor.

But although PRDC was certain everyone would find the proposition irresistible, no applause was forthcoming. Murmurs in the Detroit area indicated that many people found the idea of Fermi considerably less than compelling. In fact, some were so little taken with it that they talked of going to court to stop it.

It seems that these people were anxious about the safety features of the proposed reactor, or, rather, the lack of them. They were not convinced that the scientists and engineers constructing the plant knew enough about the behavior of atomic materials in reactors to predict how they would behave under power.

Of course, many persons who know little about reactor physics think of only one thing when someone mentions nuclear energy, and it is likely that some of those who objected to Fermi labored under the mistaken notion that an accident there could set the reactor off like an atomic bomb. Promoters of the project therefore took pains to assure the worried that such an event was, by every known law of physics, out of the question.

And yet few took much comfort in these reassurances. All they knew was that the building would house enough nuclear material to flatten dozens of Hiroshimas. That much lethal material sitting in a neat little pile thirty miles southwest of one of the world's industrial capitals simply wasn't calculated to inspire ease of mind, even if it *couldn't* go up in a mushroom cloud. Some of them had read John Hersey's book on Hiroshima, and it was difficult not to dwell on his images of the leveled city and the mutilated souls "lucky" enough to have escaped instantaneous vaporization in the fireball. These Detroiters knew it was irrational, that we had tamed the atom. Still . . .

Since these fears had no technical basis in reality, the proponents of the plant dismissed them as invalid. A good public relations campaign, it was viewed, would dispel most of the anxiety. Once people understood that an explosion was an impossibility, they'd buy the idea without a qualm. And if not—well, you can't please everyone.

What was more disturbing to PRDC, however, was the body of *enlightened* men and women who did not want the reactor built. *They* knew very well that Fermi couldn't go up in a mushroom cloud—but that was not the point. They were talking about a conventional-sized explosion, the kind that could happen if too much melted nuclear fuel came together inside the reactor core. They talked about unperfected cooling systems. They talked about potential failures of safety devices, of power failures, of unthought-out emergency procedures, of untested materials, of untried construction techniques, of unplanned radioactive-waste disposal systems, of human error. They talked in facts and figures, and they talked with authority.

And what was disturbing above everything else was that *a body of extremely knowledgeable technical authorities designated by the Atomic Energy Commission itself had deep misgivings.*

The Advisory Committee on Reactor Safeguards is an august panel of experts established by Congress for the purpose of advising the parent AEC on the safety of proposed reactors. The AEC was required to submit all data on those reactors to the ACRS for evaluation. Although the ACRS is not empowered to implement its findings or impose sanctions on applicants, its opinions should, by virtue of its members' integrity and prestige, be tantamount to law. One would think, certainly, that if the ACRS had serious reservations about the safety of a reactor, all forward progress on it would be stopped until all objections were totally satisfied.

The AEC opened the Advisory Committee's report on Fermi, dated June 6, 1956, to read:

> Although there are no facts or calculations available to the Committee that clearly indicate that the proposed reactor is not safe for this site, the Committee believes there is insufficient information available at this time to give assurance that the PRDC reactor can be operated at this site without public hazard.

It can be imagined that AEC's reaction would have been to hold up all further progress on Fermi until the information void had been filled. It must be understood, however, that in 1956 the Atomic Energy Commission's role as promoter of nuclear power was just beginning to show signs of forward movement. American industry and investment capitalists were scrutinizing the reactor program for signs of vigor, and the breeder reactor was looked to with special hope because it would mean the difference between unlimited fuel and a fuel supply largely exhausted before atomic power became established. But although the AEC must have recoiled sharply from the humiliating prospect of having to confess that Fermi might be a gross and dangerous mistake, it is almost impossible to understand what motivated the commission to do what it did next.

What the commission did, according to Representative Chet Holifield of California, was to suppress the advisory committee's negative report.

Had not one commission member, Thomas E. Murray, courageously stepped forward to disclose the AEC's action, it might never have come to light. When Murray did, the "fallout" hit the fan. The AEC's excuse, brought out later, was that the commission was under no obligation to follow the committee's advice—that it was only one of the factors to be considered along with others in reaching a decision as to whether it was proper to issue a permit. This explanation put things in an even worse light, because it raised questions about the independence of this all-important safety advisory body, and about the AEC's apparently self-contradictory dual role as both promoter of atomic energy and regulator.

Representative Holifield, by no means backward in advocacy of commercial nuclear power, termed the AEC's action "reckless and arrogant." And the indignation expressed by Senator Clinton Anderson of New Mexico, Holifield's chairman on Congress's Joint Committee on Atomic Energy, was nothing less than olympian. He called the AEC's proceedings a "star chamber" and strongly suggested that the commission had yielded to expediency in overruling the ACRS report:

> From a practical standpoint, AEC might feel obligated to go through with a bad deal with respect to public safety because they will have permitted the expenditure of huge sums under the construction permit. It is my belief that decisions on safety should be made without any examination of dollars involved, but only from the standpoint of human lives.[7]

When the AEC, even after admitting that "this fast breeder was the most hazardous of all the reactors,"[8] determined to go ahead with Fermi anyway, Senator Anderson began sketching battle plans. And when the AEC, on August 4, 1956, issued its approval of the construction permit just in time for the ground-breaking ceremony at Lagoona Beach, Anderson picked up the phone.

Among the most important persons he reached were G. Mennen Williams, then governor of Michigan, and President Walter Reuther of the United Auto Workers. He told them how the AEC's blue-ribbon panel had urged postponement of Fermi, and how the AEC had so cavalierly put the report in a drawer. He told them all he knew about the hazards of reactors, and much of what science *didn't* know about them. And although he may not exactly have told them in so many words to sue, he left them in no doubt that merely a few picket signs and baby carriages weren't going to stop this juggernaut of vested interest from triumphing.

The State of Michigan, caught in the squeeze, was to hold to a nonpartisan course in the ensuing events. But the unions dug in for a fight. At the end of August, three big labor organizations petitioned to intervene in the Atomic Energy Commission's procedures. They alleged that the construction permit "will result in the construction of a reactor which has not been found to be safe, and whose operation will create a hazard which would place Petitioners, their members and their families in danger of an explosion or other nuclear accident which would imperil their lives, health and property." Citing the reservations raised by the AEC's own advisory committee, the attorney for the petitioners stated:

> The greatest potential hazard presented lies in the possibility that the tremendous accumulation of radioactive fission products imprisoned in the fuel might somehow be released into the atmosphere and be distributed by wind, and contaminate inhabited areas. . . . *These fission products are more toxic per unit weight than any other industrially known materials by a factor of a million to a billion.*

The AEC came back with the argument that it was merely issuing to the development syndicate a permit to *build* the reactor. AEC would never issue a license to *operate* it until satisfied that all safety conditions and standards had been met. And certainly, AEC added, between commencement and conclusion of Fermi's construction the serious kinks and information gaps in the reactor's plans would be resolved.

This view was greeted with some skepticism by the unions, who reasoned that if you invest forty million dollars in construction of a plant, you darned well expect clearance to operate it when it's finished. Was the AEC prepared for the kind of pressure it would meet if it decided against issuance of an operating license after a utility had poured so much money into the project? Could rugged businessmen be expected to selflessly write off their titanic outlay in the interests of public safety if the reactor didn't pan out? The labor unions did not think so; and who knew more about rugged businessmen than labor unions?

Attention focused early in 1957 on the soon-to-be-published findings of a study team appointed by the AEC to determine what the chances of nuclear plant accidents were, and what the consequences might be. The team had been directed to think in the most pessimistic terms possible, the idea being to get some notion of the worst we could expect, or what nuclear scientists call the "maximum credible accident." But while the credentials of the group were faultless, the objectivity and pessimism with which they'd been charged somehow got subverted in preparation of the study. For when *Theoretical Possibilities and Consequences of Major Accidents in Large Nuclear Plants,* known also as the Brookhaven Report, was issued that March, its primary conclusion was that a major accident was almost beyond possibility.

The team heavily stressed that "the likelihood of accidents which would release major amounts of fission products outside the containment . . . ranged from one chance in 100,000 to one in a billion per year for each reactor." Rearranging its figures, the committee estimated that "there would be one chance in 50 million per year that a person would be killed by reactor accidents." This the AEC favorably weighed against the chance a man faces of being killed in an auto accident in the United States annually: about one in a mere five thousand.

While the desired response to these statistics may have been for Americans to abandon their autos and crowd close to the walls of the nation's nuclear plants, the effect was quite the opposite. The reason is that the AEC's study team also surmised that in the extremely unlikely event of a major reactor accident, people could be killed at distances up to fifteen miles and injured up to forty-five miles. As many as 3,400 could be killed and 43,000 injured. Property damage, resulting mainly from radioactive contamination of land, could range as high as seven *billion* dollars, the area affected being as great as 150,000 square miles.

However much the AEC might have hoped to use *Theoretical Possibilities* to pacify opposition to Fermi and other potentially controversial nuclear projects, the report backfired. The odds against a major accident were largely overlooked by the report's audiences.

All people could think about was 3,400 killed, 43,000 injured, seven billion dollars in property damage. And instead of having their questions answered, they were asking a host of others, for the report had raised a threat beyond anything Americans had ever had to contemplate short of atomic war.

The builders of Fermi, PRDC, had meanwhile turned the problem over to the Engineering Research Institute of the University of Michigan, hoping perhaps to get better odds. But this one blew back in their faces too. The institute replied in July that the number killed by escape of radioactive material from Fermi could run as high as 133,000. The report went on to scoff at the possibility of that kind of accident: "An incredible event!" it declared.[9] But what people remembered was the death toll.

More recent studies, such as the 1965 Brookhaven update, the 1974 Rasmussen report, and the 1976 Webb study (all of which will be discussed later) did not instill public confidence. Webb, for instance, extrapolated the data of the original Brookhaven study—known to the AEC as WASH-740—to account for "the sixfold increase in the highly intense, short-lived radioactivity and the fifteen-fold increase in the long-lived radioactivity" found in modern reactors, which, at up to 1300 Mw and more, are far more powerful than those planned in 1957. His results are worth quoting at some length:

> The maximum conceivable consequences of the worst accident are as follows: (1) a lethal cloud of radiation with a range of seventy-five miles and a width of one mile; (2) evacuation or severe living restrictions for a land area the size of Illinois, Indiana, and Ohio combined (120,000 square miles) lasting a year or possibly longer; and (3) *severe long-term restrictions on agriculture* due to Strontium 90 fallout *over a land area of the size of about one half of the land east of the Mississippi River (500,000 square miles)*, lasting one to several years, with dairying prohibited "for a very long time" over a 150,000 square mile area [emphasis ours].[10]

The AEC emphasized that the question of harm was largely irrelevant because of the low probabilities. Indeed, when the intervening unions sought to expand the issues in the Fermi case to include many of these crucial matters, the AEC denied their motion, interpreting the commission's rules of practice as allowing no such privilege.

Although these wranglings over potential damage and harm, and over the AEC's autocratic control over quasi-judicial proceedings, were anything but irrelevant, they were definitely at this stage academic. There had never been a reactor accident so severe that radioactive material escaped in appreciable doses from the site.

And so, as if to oblige the litigants in the Fermi action, the Number One Pile at the Windscale Works in England, a breeder reactor, suffered a serious accident on October 10, 1957, spewing fission products over so much territory that authorities had to seize all milk and growing foodstuffs in a four-hundred-square-mile area around the plant. According to Sir John Cockcroft, a leading British nuclear scientist, considerably more radioactivity was released at Windscale than is released during an explosion of a Hiroshima-type atom bomb.[11] Command Paper 302, a British report on the disaster, stated that all of the reactor's containment features had failed.

The Windscale reactor and the Fermi reactor were by no means identical, and Fermi's apologists hastened to point out that their facility possessed features that ruled out a Windscale-type accident. But their antagonists hastened to reply that Fermi was, after all, as much a one-of-a-kind proposition as Windscale had been. Could not Fermi be prone to its own unique kind of disaster? The U.S. Naval Ordnance Laboratory, in its containment study of Fermi, was to state: "Many of the containment problems that are peculiar to this reactor have never been solved experimentally or theoretically."[12]

Leo Goodman, speaking for the United Auto Workers, put it in graphic terms that people in the Motor City could appreciate: "To locate this experimental plant that has never been brought to full power so near population centers is as reasonable as trying to control a ten-ton truck with untested brakes in a congested city street," he said.[13]

Of course, it *was* reasonable in one respect. Because it is expensive to transmit electricity over long distances, utilities find it expedient to build power plants as close to population centers as possible.

Goodman's simile, and everything else the unions threw at PRDC-AEC, was to no avail. On December 10, 1958, the commission, with a few modifications, reaffirmed its original permit to PRDC, and on May 26, 1959, issued its "Opinion and Final Decision," dismissing all exceptions taken by the petitioners on the grounds that there was enough assurance, for the purpose of the provisional permit, that Fermi could be built and operated without risk to the health and safety of the public.

But the unions were not to be put down so easily, and they went to court. There their assertions were upheld. Two of the three circuit judges hearing the case in the U.S. Court of Appeals, in a decision rendered June 10, 1960, felt the AEC's permit should be set aside. The AEC's "predictions," the decision stated, did not satisfy the requirements of the Atomic Energy Act of 1954 *that safety be established before, not during or after*, construction gets under way. "The possibilities of harm," Judge Edgerton said for the majority, "are so enormous that any doubt as to what the Act requires, and any doubt as to whether the Commission made such findings, should be resolved on the side of safety."

The AEC and the Fermi people did not accept this line of reasoning, however, especially as construction of the plant was already far along. So they carried their case to the Supreme Court, on the grounds that the lower court's decision would have a "seriously disturbing effect on the development of the peaceful uses of atomic energy"—a thoroughly irrelevant argument.

While the highest court in the land was considering the merits of the case, the lid blew off the SL-I reactor in Idaho Falls, killing the three men tending it.

Unfortunately, the implications failed to register on judicial geiger counters. The majority of justices on the Supreme Court upheld the PRDC permit, and Justice Brennan, delivering the opinion on June 12, 1961, affirmed his faith in the AEC to stand up against business pressures when the time came to issue an operator's license. "PRDC has been on notice long since that it proceeds with construction at its own risk, and that all its funds may go for naught. With its eyes open, PRDC has willingly accepted that risk, however great."

Justice Douglas, with Justice Black concurring, took the most strenuous issue with the verdict, declaring: "The construction given the [Atomic Energy] Act by the Commission is, with all deference, a light-hearted approach to the most awesome, the most deadly, the most dangerous process that man has ever conceived."

But that, as the court had found, was not relevant.

The legal issues resolved at last, PRDC finished construction and began testing Fermi's systems. But what the laws of man had put together, the laws of nature now conspired to tear asunder. From the outset the plant seemed jinxed. According to Saul Friedman, a *Detroit Free Press* staff writer reviewing the reactor's history up to July 1966, nothing worked the way it was supposed to. The thirty-inch-long pins containing uranium fuel were designed to last until 3 percent of each pin's fuel had been consumed. But after a short time they swelled, making it difficult for the liquid sodium coolant to flow between them. So the engineers had to replace the fuel core after only four-tenths of 1 percent of the fuel had been consumed.

You don't replace fuel cores the way you change tires. The process is so delicate and complicated that it took a year to make the switch—the bill coming to four million dollars. Then, when they finally got the new core in, they found that liquid sodium had clogged a number of key components. It had also eroded some graphite, a radiation absorber used to "moderate" or control the fuel's activity. That meant more graphite.

Then, in Friedman's words:

> The steel dome over the reactor vessel had to be redesigned to prevent the giant plug that bottled up the reactor from shooting through the roof in the event of explosion.
> The mechanism to move the fuel elements in the reactor failed. . . .
> The giant cask car, which transports radioactive elements, failed. . . . The giant sodium steam generators constantly caused trouble, and one nearly exploded. They leaked sodium.

Some of the technicians wondered if the reactor was trying to tell them something.

By the summer of 1966 it was estimated that Fermi, originally financed for between $40 and $45 million, had cost about $120 million. It had produced no more than $303,000 worth of electricity and bred nary a gram of fuel. The reactor hadn't been brought to anything near its maximum capacity, and the way things were going it would be a long time before it even approached it—to say nothing of sustaining it.

In spite of all these setbacks, the reactor still bumbled along, trying to work out the $80 million worth of bugs it had developed during its ten-year sojourn. But although a lot of observers were beginning to call this golden goose a white elephant, it still had one more egg to lay.

And that brings us up to October 5, 1966.

A year after the accident, it was announced that the eight-inch piece of metal in Fermi's core, pinpointed as the immediate cause of the accident—and it was *not* a beer can, as had been suggested half-facetiously by some—had been identified. According to *Scientist and Citizen,* it was one of six identical sheathing elements in the cooling system, elements not even called for in the original plans for the reactor nor officially recorded after installation.[14] They'd been thrown in, you might say, as last-moment safety measures during construction. Unfortunately, a workman had failed to secure one properly, and it had been swept up by the sodium rushing through the cooling system and cast against coolant nozzles, blocking them. This caused several fuel subassemblies to overheat, warp, melt, and shove several more out of kilter, making the meltdown, in the words of Sheldon Novick, reporting in *Scientist and Citizen,* "a bit worse than the 'maximum credible accident.'"

There was a lesson in the experience, and AEC, taking a cautious stance toward the future of the breeder program, seemed to be willing to learn it. Not so with the business community. The Fermi reactor was cool but a few months when, at a panel discussion of the Atomic Industrial Forum's winter convention, Chauncey Starr of Atomics International complained: "The national fast-breeder program is too small, too slow, and too timid."[15]

Still troubled by mishap after mishap, Fermi was started up again four years later. It ran, balkingly, from July 1970 to August 1972; it was then permanently closed.[16]

2

Those Who Favor Fire[1]

President Truman, reeling from the horrors he had authorized to be unleashed on Japan, described atomic energy as "a force too revolutionary to consider in the framework of old ideas." It stood to reason, then, as Congress took up the burden of postwar legislation of atomic energy, that if the energies involved were nothing less than cosmic, and the human prospects and national destinies nothing short of fundamental, then the responsibilities to be fixed on those designated to regulate atomic energy would be close to herculean. After all, someone—a man or a group of men—had to be entrusted with control over forces directly bearing on human survival. The leverage inherent in that trust was practically infinite; the wisdom and restraint involved would demand the best of those men.

Many Americans were overwhelmed with remorse and fear over the awful destructive energy we had liberated. Sentiment ran high in many quarters for the total abandonment of any kind of atomic power, for good or for evil. Humanity had gone too far, they were saying, and must turn back at once before it was propelled on a mad, irrevocable course.

Tragically, it was out of the question. Russia's dismaying belligerence, and its ominous capture of some German nuclear scientists, ruled out any possibility of dropping our guard, let alone our ultimate punch. But even if Russia had not so behaved, it was naive to think our government would dump its atomic bombs and all pertinent information into the sea. Sooner or later, someone else would learn how to make A-bombs; we were obliged to maintain our advantage.

One fact, therefore, was inescapable: This nation was committed to continuation of research and development of atomic energy, particularly for the purpose of defense. It was a fact that many must have found unbearably bitter. Indeed, it would not be hyperbole to suggest that immediate rearmament, and rearmament with atomic weapons following the most savage war of all time, was a fact the American nation, as a psychological entity, literally could not face. The traumatic cruelty of it demanded some sanity-saving rationale, some desperate hope to which we could cling while we stockpiled atomic bombs. That hope took the form of a belief in the atom's capability for peaceful service. Many psychologists and social scientists have advanced this theory, stating that the American people as a whole, in order to assuage their guilt feelings over the destruction of Hiroshima and Nagasaki—and, more important, over the godlike (or demonlike) powers their science had created—embraced atomic energy as an important contributor to the peacetime needs of the world's people.

This conviction was given great impetus by many scientists who— perhaps for the very same reason of compensation for guilt—sincerely believed that the atom could be turned from mankind's foe to one of its closest allies, that it could be clean, safe, reliable, and economical. Psychology aside, a great deal of scientific optimism about the future of atomic power was founded on a gross misjudgment of the problems and dangers inherent in the young technology. In the mid-1940s, and indeed well into the 1950s, many of the most serious reactor safety problems were unknown. Most reactor malfunctions were completely unexpected, since, as Dr. James McDonald has pointed out, they would have been engineered out had they been anticipated.[2] Barry Commoner in an article in *Scientist and Citizen* has described how the biological and genetic effects of different radioisotopes were a very long time in being recognized; it was not until weapons testing had already introduced quantities of these poisons into the environment that short- and long-term consequences could be described with reasonable scientific accuracy.[3]

These then were the factors that accounted for America's embarkation on the journey into peacetime nuclear power. The need to continue development of atomic weaponry provided the technological impetus; the dream of the atom's commercial and humanitarian usefulness—a dream fostered by what later proved unjustified optimism on the part of many influential nuclear scientists—provided the rationale. And pervasive guilt about stealing divine fire supplied the psychological climate in which the whole mammoth venture could thrive.

Congress and the President, no less than the American people—and perhaps more so, in view of their direct role in the development and employment of the atomic bomb—were subject to these psychological fluxes and found the vision of a future energized by the peaceful atom difficult to resist. These were the circumstances in which the Atomic Energy Act of 1946,[4] setting up the program and providing for the establishment of the civilian-run Atomic Energy Commission, was passed.

But Congress, an institution that was supposed to know something about political power, still could not or would not cope with the fact that unprecedented responsibilities demand unprecedented prerogatives. Perhaps no legislation could have been framed so as to curtail the AEC's potential privileges. Whether that is so or not, the Act created opportunities for the assumption and exertion of prodigious, if not unlimited, power.

Two major faults were built into the 1946 Act, either of which would have been grave enough by itself in a matter of such overriding importance, but which together constituted a truly formidable governmental force. The first of these was the conferment of unprecedented power on the Atomic Energy Commission; the second was the granting to the commission of unprecedented independence and privileges of self-regulation.

As to the first, the Act stipulated that the AEC would have exclusive control over production, ownership, and use of fissionable material. Furthermore, the control extended totally or largely into such areas as mining and refining of ore, research and development, information dissemination, radiological health and safety programs, licenses and agreements, bomb production and other military applications, patents and inventions, security matters, administration of international arrangements on atomic energy, the right of eminent domain over lands containing radioactive resources or lands on which atomic facilities were to be built, and of course ownership of the facilities themselves.

The establishment of a government monopoly shattered what Truman had called "the framework of old ideas," and the Act of 1946 made the AEC exclusive steward of the new framework. Some reasoned that such a monopoly was necessary then in view of the delicacy and secrecy involved. That may have been so at the outset, but time was soon to alter the conditions in which atomic energy operated, making the AEC's immense power a dangerous archaism. Senator Thruston B. Morton, in 1968, reminded his colleagues of "the oft repeated statement in those days that electrical power from atomic reactors would be so cheap that it would not even be worthwhile to meter it," and of the now ironic words of General Leslie Groves when he turned over nuclear responsibility for the Manhattan Project to the Atomic Energy Commission: "You of the Army's Manhattan Project . . . have raised the curtain on vistas of a new world."

"No one," continued Morton, "questions that the development of the ability to create electrical energy through atomic fission is a tremendous accomplishment, or that someday in the distant future we may be forced to depend on it after our other bountiful sources of electrical energy are exhausted or become too scarce and costly to utilize. But we also know that atomic energy is not the panacea of all our energy problems it was once expected to be, and we are becoming more aware every day of the costs in terms of potential danger to humanity which this proliferating atomic energy program may entail."

The second major weakness in the original Atomic Energy Act was the extraordinary independence it afforded the commission. Part of the problem was that the commission was *a* commission, because commissions in our government are by nature free of much of the restraint that characterizes other governmental bodies. These regulatory bodies "drift along," in the words of former AEC member Thomas E. Murray, "somewhere in a 'twilight zone' among the three branches of the Government—President, Congress, and the courts. Subject to all three branches in specific respects, the commissions nevertheless evade complete and continuing control by any one of them. That is why they have sometimes been called the 'headless fourth branch' of the Government."[5]

Thus commissions of *any* sort, being neither fish nor fowl, can with any degree of power create for themselves a most advantageous niche from which to dictate without being dictated to. But what is one to say about a commission endowed with *unprecedented* powers *and* "drifting along somewhere in a 'twilight zone'" among the three branches of government? Was this not a unique combination? Would the commission, as time went by, be able to resist taking advantage of its dizzyingly special status?

It was not simply that there were too many loose threads; it was that the threads trailed from a mantle that few men in government had ever worn.

On paper the AEC appeared anything but independent, for its activities came under the jurisdiction of all three branches of government. The President of the United States had the power to hire, with the advice and consent of the Senate, the five members of the commission, and he could remove them before expiration of their terms for "inefficiency, neglect of duty, or malfeasance in office." The President also exercised general executive power over all independent AEC commissions, had considerable control over the budget, over classified information, over weapons quotas, and over settlement of disputes involving the AEC; and the President had to approve all agreements for international cooperation in atomic energy. Furthermore, the Act called for establishment of a Joint Congressional Committee on Atomic Energy to review all proposed legislation pertinent to atomic energy, to hold hearings and make recommendations to Congress. Finally, judicial review of the commission's actions was provided for by the Act.

A number of factors, however, balanced and eventually outweighed these government controls. The first was the scientific complexity, indeed the sheer opacity, of atomic energy as far as the layman was concerned. The nature of the work was far beyond the technical comprehension of men in other branches of the government. Only scientific experts understood it, or said they did, and everyone else from the President—whose scientific advisers were largely pro-atom —on down had to take it on faith. Second was the top priority of maintaining a big lead over the Russians in atomic research. National survival was at stake, and our government had to give the atomic energy establishment the widest latitude in the interests of national security. Third was the rapid proliferation of a bureaucracy, replete with Advisory Committees on This, Divisions of That, Offices of the

Other Thing, panels, boards, making it more difficult for government overseers to keep track of what the commission was doing, and therefore it was more difficult to maintain control.

Here then were all the ingredients for a virtually independent government within the government, a scientocracy invested with nearly complete powers of policy-making and self-regulation. Furthermore, the Atomic Energy Commission found itself the repository of a dual and conflicting role: It had to *promote* an atomic energy program on the one hand and *regulate* it on the other. The position was not unlike that of the family that owns the sole grocery store in town and serves as mayor, sheriff, judge, and town council as well. Getting one's way becomes a simple matter of donning the right hat.

Many years afterward, the commission's first chairman and later a harsh critic of atomic policy, David E. Lilienthal, was to score the AEC's best-of-both-possible-worlds position and the danger it posed to the public:

> It is unfortunate that the AEC is not only the overall protagonist of a nation-wide atomic-power program; it is also the body that must sit as judge of the safety to the public of the design, mode of construction, and site of particular atomic power plants. In short, the AEC, as a general promoter of atomic power, must also decide the quasi-judicial issue of whether a license is issued. With a world of goodwill and integrity and technical competence on the part of the AEC, how well is the public protected by this dual and conflicting role?[6]

The mischief inherent in the situation did not become apparent until the early 1950s. Certainly, the original members of the commission conducted themselves with the highest probity. (It should be emphasized here that the integrity of the AEC is nowhere in this book called into question: only, to use David Lilienthal's distinction, its infallibility.) The commission of the late 1940s sincerely believed that in promoting this new discovery for man's good, it would be rendering an invaluable service to the country. Furthermore, during that period the commission did not, any more than anyone else, fully appreciate the complications and dangers residing in a commercial reactor program, nor did it completely recognize the extent of the responsibility that had been placed on its members' all-too-human shoulders. Undoubtedly it *was* conscious of the duality of that responsibility, however, and the conflict between promotional and regulatory must have caused many sleepless nights for the conscientious personnel of the AEC.

During this period, roughly to 1953, the foundations of power reactor technology were laid. Basic research into properties of various fuels, coolants, and moderators was undertaken, designs and materials were tested, experimental reactors were built. While electricity generation was an important goal, most of the allocations, labor, and practical knowledge went into development of reactors for propelling submarines and aircraft. As long as military priorities prevailed, the AEC could not make its debut in the marketplace.

But military priorities could not prevail forever. The airplane reactor technology flopped, and naval reactor technology did not promise to absorb anywhere near our total nuclear effort. Neither did nuclear armament. Even though the Russians had the atomic bomb, and though nuclear fusion, the principle on which the hydrogen bomb is based, had given the Cold War a new boost, the arms race did not strain the capacity of the atomic energy program. And of course, the public was growing eager to see the atom put to uses other than belligerent ones, to see it fulfill the promise envisioned in the days right after the war. Russia was building a nuclear power plant that would be ready before we got our own power program going; *that* was incentive enough right there.

Looking back to the early 1950s, our government must have realized it had an enormous investment to protect, not the least part of which was prestige in the world community. Looking ahead, it could foresee its titanic atomic industrial effort bogging down unless it broadened its base and went commercial. The Atomic Energy Commission therefore, cautiously at first, began to remove the hat of self-restraint and put on the hat of promoter.

Though nuclear information was still highly classified, some non-priority data was cleared, and a number of qualified industrial teams were invited in 1952–53 to study it with a view to evaluating the economic outlook. A number of them came away smiling. For one thing, the AEC itself was buoyant about prospects, the threat of major technological headaches being at that time a cloud no bigger than a man's hand. For another, the visiting teams may have glimpsed stimulating incentives. It is generally acknowledged that the government gave assurances to the effect that the plutonium produced as a by-product of the fission process would be bought back by the government for good prices—plutonium being essential to atomic weapons, of course. Saul Friedman, a *Detroit Free Press* staff writer, in an article about the group that pursued the breeder reactor franchise, suggested that the AEC had encouraged them to believe the government would pay them ninety dollars a gram for their plutonium, which was almost three times as much as they were paying for high-grade, weapons-type plutonium.[7]

It is also possible that the industrialists who peeked into the AEC's cupped palms saw something rather distressing, in the form of a hint that if private capital didn't develop a nuclear power industry, Uncle Sam would extend his atomic monopoly to include the generation of electric power, in direct competition with the utilities. The smiles on their faces must have frozen when they saw that.[8] In any event, the consensus of the industry observer teams was that nuclear plants could produce electricity at a cost roughly competitive with conventionally fueled plants.

Heartened by industry's optimistic if totally predictable response, the AEC now set the hat marked "promotion" at a more rakish angle, preparing a breathtakingly ambitious blueprint for the advent of commercial nuclear power. Some officials were talking about half of America's electricity being generated by the atom by the end of the century.

Of course, the Atomic Energy Act of 1946 prohibited private initiative in the nuclear power field, and it was therefore necessary to effect a change in the law. Since that change had the official approval of the President, the support of the Joint Congressional Committee on Atomic Energy, and the apparent approval of the electorate, the conclusion was foregone. Congressional opponents, armed only with their nameless dread, were pushed aside easily. In 1954 the Atomic Energy Act was rewritten to permit private organizations to build and own atomic energy facilities and operate them under AEC license and regulation.

The rationale for this effort was termed "Atoms for Peace," a truly inspiring motto, and it had been enunciated by President Eisenhower before the United Nations on December 8, 1953: "The United States pledges before you—and, therefore, before the world—its determination to help solve the fearful atomic dilemma—to devote its entire heart and mind to find the way by which the miraculous inventiveness of man shall not be dedicated to his death, but consecrated to his life."

As recently released documents reveal, however, Eisenhower himself had major qualms about the atom and its uses. In discussions with then AEC Chairman Gordon Dean on the atomic bomb testing in Nevada (cancers from which have begun to appear in large numbers only recently), Eisenhower "expressed some concern" on May 27, 1953. According to Dean's diary, the President went on to suggest that "we leave 'thermonuclear' out of press releases and speeches. Also 'fusion' and 'hydrogen.'" Dean's conclusion: "The President says 'keep them confused about fission and fusion.'"[9]

The rhetoric of Atoms for Peace was dangerously simplistic. The mellifluous phrase set a classic pattern for the glib promotional slogans for atomic energy that were soon to follow, in which the atom became a good friend and neighbor, a plowshare beaten from swords. And into these hollow catch phrases crept the notion that radioactivity consecrated to the benefit of mankind was somehow less poisonous than that dedicated to man's destruction. Battening on mankind's profoundest yearnings, the notion quickly swelled into a widely held assumption, and it was eventually adorned by promoters with the trappings of gospel.

Not everyone was carried away by these tantalizing vistas. Representative John P. Saylor of Pennsylvania, speaking before Congress some time later, was to recall the deep misgivings he and many other colleagues felt about the great leap forward into commercial nuclear power:

It was not too many years ago, Mr. Speaker, that the general public was excited at the prospect of the development of an atomic reactor which purportedly would bring vast cost savings to consumers of electric power. It is true that a number of Members of Congress had serious doubts about the practicability of using millions upon millions of dollars of U.S. Treasury funds merely to utilize a new source of energy for power generation. I recall that many of us stood on the floor of this House time and again as far back as 1956 to question the wisdom of such tremendous expenditures, particularly when the U.S. Geological Survey has established without qualification the existence of sufficient coal reserves to satisfy the power requirements of the entire nation for at least a century to come. Congress also was reminded of warnings by distinguished scientists who believed that the safety

issues entwined in the fission process should be resolved before the wholesale construction of nuclear facilities.

But caution and economy could not prevail in a climate of optimism
and enthusiasm generated by wanton promises of miracle and magic
through applications of a glamorous rare element whose material value had
previously been confined exclusively to its destructibility factor.

Such profound reservations, however, carried no weight against
an extremely determined government policy reinforced by the clamor
of a utility lobby eager to get in on the ground floor of potentially the
most prodigally subsidized undertaking in American peacetime history. The Atomic Energy Commission, of course, was to be the instrument of that policy, and it was around this time that the complexion
of the commission began to change visibly. The commission began to
emerge from its relatively passive role to become a most aggressive
instrument indeed. It started picking up the loose threads that the
Atomic Energy Act of 1946—and to an even greater extent that of
1954—had left hanging, and it began to feel at ease in the mantle
from which they were suspended.

Unfortunately, at the very same time, from 1954 onward, serious
problems began cropping up in reactor technology, and bomb tests
disclosed somatic, genetic, and environmental dangers far worse than
anyone had anticipated. So now the AEC had a double problem; it not
only had to promote atomic power against ordinary resistance to anything new, it also had to promote it against concrete evidence that it
might not be safe, might not be clean, might not be reliable, might not
be economical. The commission probably could not see it clearly, for
it all happened gradually and subtly, but the only way out of the bind,
aside from admitting that we had all made a gigantic mistake, was to
begin erecting an elaborate structure of psychological defenses.

Although a large number of companies came into the AEC's camp
directly after passage of the 1954 Act, it would be wrong to assume
that private industry, uniformly, was deliriously happy about the new
status of atomic energy. Naturally, component manufacturers, utility
operators, and other industrialists and businessmen followed these
developments with keen interest, but when it came to a deeper commitment, many dragged their heels. It was not that they doubted that
American know-how could lick the technical problems in due time.
But from an investor's point of view, nuclear technology was still very

much in the gestation state, and the economic challenge was far less surmountable than was being generally and generously proclaimed. The hazards were still a giant question mark, and questions of liability had scarcely been thrashed out. Why not wait and see how these questions were resolved, and how the experimental and demonstration projects worked out before casting one's lot with Atoms for Peace?

For six months following the 1954 Act, therefore, the AEC found itself in the position of a puzzled hostess on the night of her party who, when no guests appear, suddenly wonders if she put the correct message on the invitations. It is true that planning nuclear power projects takes time, especially when much data is classified, as was the case in 1954. Nevertheless, industry seemed to be shuffling its feet, waiting to see how the government's own pioneer reactor, the 60,000-kilowatt plant going up at Shippingport, Pennsylvania, would do. Not a single application for a construction license came in. "There is little question but that industry's initial response was disappointing to a Joint Committee and an AEC bent on accelerating the national nuclear power effort," says John F. Hogerton in a *Scientific American* review of the growth of atomic power.[11]

By winter the AEC realized there was need to prime the pump, and accordingly, in January 1955, the commission announced a "Power Demonstration" program designed to stimulate plant construction. Liberal aid was offered to any utility prepared to put up a plant, in the form of research and development assistance and waiver of fuel inventory charges for the first five years of plant operation. At last, that March, the first application came in, and by year's end there were two more takers. In addition, two other projects, privately financed rather than plugged into the Power Demonstration program, were launched.

Despite this forward motion, industry as a whole continued to hesitate, and it steadily grew apparent that the real reason why many businessmen had supported the 1954 Act was to ward off a federal government threat to go into the electricity business if private industry declined to take the nuclear initiative. James W. Kuhn, in *Scientific and Managerial Manpower in Nuclear Industry,* confirms this view:

> Under the threat of public power, several utility companies did come forth with plans for large-scale nuclear power plants. Officers of the companies expected no profit from the plants. The incentive was largely negative—to

keep civilian nuclear power private. As the president of one of the companies remarked, "We acted because we needed to guarantee the position of private industry. The money spent was a gamble to preserve the private sector." The president of another of the companies explained why his management pushed into nuclear power: "We made a proposal on what became Shippingport and we breathed a good deal easier when we didn't get the contract. We weren't anxious to get into nuclear power, and I don't think any other company in its right mind wanted to get into it either. But you see, we had to bid—we had to act—whether we wanted to or not. We had been pushing the private development of nuclear power and we couldn't refuse to get into it after pushing so hard."[12]

The lobby had managed to get a clause written into the 1954 Act specifically forbidding the government to engage in the sale or distribution of electricity for commercial use. Now that the threat was averted, and the lobby had it in writing, private industry could take its time going nuclear. That clause had a big loophole, however: It permitted Uncle Sam to sell electricity "incident to the operation of research and development facilities of the Commission." What was to prevent the government from building scores of "research and development" facilities and selling the "incident" electricity all over the country? This was one of the AEC's bigger whips, and the commission applied it adroitly.

But as all mule skinners know, a carrot ahead is worth a whip behind. When the power industry in 1955 continued hedging, taking just enough initiative to keep the government happy, and minimizing risks by forming large joint-venture investor groups, the AEC announced the "second round" of its Power Demonstration program—more subsidies. Three more reactors went up as a result, but *still* the industry balked.

Late in 1956 the Argonne National Laboratory's experimental boiling water reactor, designed specifically for electricity generation, went into operation, and the next year Shippingport was completed. Industry nodded appreciatively, but its attention was really fixed on the units being erected by the utilities, not on those by the government. The former were proving more expensive to build than expected, and it looked as if they might prove to be more expensive to run as well. And the fossil-fuel industry was preparing a counterattack that would make it much tougher for the government to produce nuclear-generated electricity at competitive prices.

Furthermore, anxiety about safety was beginning to percolate. There had been talk at the 1955 Geneva conferences on atomic energy about the probabilities and consequences of a major accident, and a number of actual accidents in government reactors had made many people stop and ponder. The issue reached a boil with the publication of the Brookhaven Report early in 1957, with its estimates of as many as 3,400 fatalities, 43,000 injuries, and $7 billion in property damage.

Would the public stand for the idea of nuclear powerhouses virtually in its backyard? Would the costs of building safe reactors make nuclear power financially unfeasible? Wouldn't the cost of insurance alone price reactors out of the market? If not, who was going to be liable for damage claims?

Apparently, until the investor's mind was set at ease on the big question of safety and liability, investment capital was never going to flow freely. Accordingly, our government held out an even bigger carrot. Perhaps "plum" describes it better, for through passage of the Price-Anderson Act,[13] it was guaranteed that private industry would not be held liable for more than a token of damage costs in the case of a major nuclear plant accident. This Act, passed in 1957, set a limit of $560 million worth of indemnity on any radiological accident, but only $60 million of that amount would be put up by private operators. *There was no fiscal responsibility beyond $560 million, even though damages could run as high as $7 billion!* Total coverage amounted to only 8 percent of potential claims, and private industry's liability came to less than 1 percent.

Although the taxpayers' share of Price-Anderson coverage has been steadily decreasing, private insurers today are still responsible for only a small fraction of the liability. The total dollar amount has never been increased, despite both tremendous increases in the size — and thus increased danger — of nuclear reactors and runaway inflation, which has reduced the buying power of the dollar by more than 50 percent in the past twenty years. While Price-Anderson was originally designed as a short-term subsidy to get the nuclear industry on its feet, it has been repeatedly extended; it will now expire on August 1, 1987.[14]

In the words of James W. Kuhn: "The chairman of the AEC, Lewis L. Strauss, made explicit the choices before industry in 1957. 'It is the Commission's policy,' he said, 'to give industry the first opportunity to undertake the construction of power reactors. However, if industry does not, within a reasonable period of time, undertake to build types of reactors which are considered promising, the Commission will take steps to build the reactors on its own initiative.'"[15]

Meanwhile, in the Fermi action the AEC had overruled the plea of its Advisory Committee on Reactor Safeguards for caution, initiating a policy of approving reactor applications before safeguard plans and technology supported such an action. Its handling of the Fermi hearings indicated a dangerous tendency to manipulate procedures in order to minimize opposition. The Atomic Energy Commission was flexing its muscles—muscles that Congress had molded in 1946 and 1954—at every sector involved in atomic energy: the public, Congress, the judiciary, the President, private industry, and the commission's own advisory bodies. The Fermi case was not to be resolved for some years, though when it finally did come to a decision, the AEC would prevail.

Despite the foregoing, however, all the plans and programs, ultimatums, and incentives could not bring into the world a technology whose limbs and vital organs were still largely unformed. With one exception, the plants built during the first and second rounds of the Power Demonstration program had weighed in at higher costs than anticipated, and one was nearly double the estimate. The generating costs achieved in these plants at the outset were 50 percent higher than predicted. Conventionally fueled power was effecting savings in a variety of ways, and the price of coal was dropping. The insurance issue was by no means resolved, for utilities knew that even if they were only nominally liable on paper, a major accident would nevertheless be ruinous. The Fermi case, on which so much was riding, began to proceed through the courts.

In the late 1950s, therefore, the commission initiated a Ten-Year Program, establishing short- and long-term goals in atomic energy. Even *then* industry stayed away: Only four more plants went up as a result of the "third round" of the Power Demonstration program. The AEC may have been demonstrating power, but it was not demonstrating might. A crisis began to loom, and after President Kennedy took office he requested a thorough assessment of the past, present, and future of atomic energy from the AEC.

The commission responded with its 1962 Report to the President on Civilian Nuclear Power, a major policy statement. Because the chips were down, this document had to be carefully worded—and it was. How, for instance, was the commission to explain its failures? In one key passage, that question was answered:

> Nuclear electric power has been shown to be *technically* feasible, indeed, readily achieved. Power reactors can be reliably and safely operated. However, contrary to earlier optimism, the economic requirements have led to many problems—combining low capital cost with long life and assured reliability; lowering costs by improved efficiency; developing long-lived and, therefore, economic fuels. Attempts to optimize the economics by working on the outer fringes of technical experience, together with the difficulties always experienced in a new and rapidly advancing technology, have led to many disappointments and frustrations. Experiments have not always worked as planned. Many construction projects have experienced delays and financial overruns. Such difficulties led to considerable diminution of the earlier optimism regarding the early utilization of nuclear power, which in turn contributed to the withdrawal of some equipment and component manufacturers from the field.[16]

However cleverly the AEC phrased it, the meaning of this admission was certain: that power reactors could not at that stage be made both technically feasible *and* economical. If we wanted safe, reliable reactors under present technology, we would have to build and operate them at a loss; if we wanted economically competitive reactors, we would have to compromise safeguards. Phrased another way: Nuclear power works in theory, and in limited experiments, but as for commercial viability it has proved to be a great big dud.

But that is not the kind of admission one makes to one's Congress and President after spending more than a billion dollars in Treasury funds. "To have second thoughts in the pursuit of a will-of-the-wisp in public programs or in private enterprise takes moral courage and aroused taxpayers or stockholders," David Lilienthal was to say the following year. And again: "The initial goal that was the justification both for the unique status given the Peaceful Atom and the gargantuan scale of public expenditure has long since proved to be a mirage. That myth of a revolution continues to be fed by the American taxpayer. Private firms repeat this fiction at this late day, as part of institutional promotion of the sale of their atomic reactors."[17]

But the AEC was not so objective. The habit of "denial," of constantly belittling hazards and problems while emphasizing progress and promise, apparently had become second nature to proponents of atomic power. And *that* was the most treacherous fact of all.

Under that circumstance, what course was there but to plunge even more deeply into the morass? In the very same report to President Kennedy that stated "for safety reasons, prudence now dictates placing large reactors fairly far away from population centers," the commission announced its decision to encourage the launching of huge commercial reactors, far bigger than anything tried thus far:

> To encourage construction of full-scale power installations by utilities, the support of research and development and the temporary waiver of fuel charges have recently been augmented by the offer of reimbursement of design costs for fuel installations of 400 megawatts [400,000 kilowatts] or more. Both public and investor-owned utilities are eligible. It is hoped that these forms of assistance will suffice to bring about a marked increase in the number of full-scale installations.

Presumably the AEC was hoping that by the time the big installations went up technology would have caught up and discovered a way to make them work. This topsy-turvy philosophy was sanctioned now by the highest court in the land, for in June 1961 the Supreme Court upheld the developers of the Fermi project.

Thus without ever having proved that power reactor technology was technically reliable, commercially viable, or safe, the AEC was encouraging industry to build reactors of untested size and immeasurable danger on sites close to major population centers.

But this encouragement apparently did the trick. The power brokers' defenses began crumbling, slowly at first, but as the mid-1960s dawned, with greater and greater speed. Private industry began tooling up in a big way. Connecticut Yankee Atomic Power Co. put in for a $13,195,000 subsidy for a 490,000-kilowatt reactor at Haddam Neck on the Connecticut River; Southern California Edison and San Diego Gas & Electric asked for $13,022,000 to build a 395,000-kilowatt reactor near San Diego; the City of Los Angeles requested $16,200,000 for help in putting up a 490,000-kilowatt reactor in the Rancho-Malibu area; Niagara-Mohawk Power filed for a reactor on Lake Ontario, although it graciously declined a subsidy, wanting only free nuclear fuel for its first five years of operation. Jersey Central proposed a 500,000-kilowatt reactor near Toms River, New Jersey: the Oyster Creek Plant.[18] With financial assistance under the AEC's Power Demonstration program, announced Jersey Central, its plant could be made competitive with a fossil-fueled plant. ("By the same token," said Representative

John Saylor in an excoriating speech in the House, "Jersey Central may have pointed out that old newspapers or imported Swedish timber would be competitive with coal if Government subsidies were sufficient to absorb the cost differences.")[19]

Another passage from Saylor's speech sums things up splendidly:

> In other words, Mr. Speaker, although at least 1.3 billion dollars has been poured into the civilian reactor program by the AEC in the past nine years, the imbalance of costs between electricity generated by the atom and that generated by conventional fuels is still of such magnitude that even the current multi-million-dollar bestowals by the Federal Government to the investor-owned utilities may have to be increased in order to get the program moving at the rate desired by the AEC. Meanwhile, irrespective of the expensive research carried out since 1954, the degree of danger hovering over a nuclear power plant remains a mystery.

By 1966 industry's resistance to nuclear power had effectively collapsed. The AEC's Annual Report to Congress for that year declared: "The year 1966 saw atomic energy become a major factor in the planning for meeting the Nation's future electric power needs as 55 percent of the new steam-electric generating capacity announced by U.S. utilities was for nuclear plants." Applications for sixteen reactors were filed that year. In 1967 the AEC's Annual Report blared that "utility planning announcements doubled the previous pace-setting growth of 1966," with twenty-nine permit applications in hand.

But the upshot of this story is a monumental irony. For by 1967 the Atomic Energy Commission, deluged by applications, pressured maddeningly by utilities to clear licenses, besieged by industry for guidance in standards and criteria, critically hampered by manpower shortages, and inundated by innumerable problems never anticipated when it put its promotion into high gear, began to grow conscious that the harvest it had so assiduously cultivated was threatening to overwhelm it.

3

Thresholds of Agony

Because much of what follows deals with the perils of radioactivity, it is important to review some of the ways in which radioactivity kills and injures, because the mismanagement of the peaceful uses of atomic energy will subject humans to many of the same gruesome afflictions and agonizing deaths as those suffered by survivors of Hiroshima's fireball. It will also create profound environmental disorders.

Radiation damage may be inflicted in a number of ways: in large doses or small, all at once or by stages, externally or internally. Irradiation may be of the whole body, of specific organs or tissues, or of only a few cells.

To give this discussion a context, one should first understand that there is and always has been an external and internal "background" of natural radiation. Until the Atomic Age, that background was constant. Experts calculate that the sum of internal and external natural background radiation is about five roentgens over the first thirty years of one's life. Consider the consequences of an intense single dose of radiation suffered by the whole body, since in its simple graphic horror it is easiest to comprehend. In the event of a major reactor accident in which radioactive gas and fission products are dispersed into the atmosphere, a large number of people will be exposed directly to massive doses of radiation. What will happen to them? While the amount of radiation necessary to produce various effects might vary according to a variety of conditions, the following thresholds expressed in roentgens can be taken as fairly universal:

A dose of 600 r or more would probably kill nearly everyone exposed within a month.

A dose of 400–450 r would be deadly in half the cases.

A dose of 300 r would kill one quarter of those exposed and induce serious injury in 90 percent of the remaining number.

A dose of 200 r would kill about 2 percent and induce serious illness in half of the remaining cases.

From 200 r down to zero the effects diminish proportionately, and between 0 and 25 r no observable effects are produced directly. But this statement must not be misinterpreted to mean that such low doses are harmless.

On the contrary, many leading scientists, such as John Gofman, Helen Caldicott, Edward Martell, and Ernest Sternglass, have claimed that low doses of radioactivity will actually lead to more cancers; at higher doses cells will be killed outright and will be unable to form tumors.

Sternglass documented measurable increases of radioactivity in milk after a 2,970,000-curie radioactivity release from the Millstone, Connecticut, plant in 1975. The levels—56 to 61 picocuries—were two to three times higher than those recorded in the area at the height of nuclear bomb testing. This, according to Sternglass, drastically changed the cancer rate in Rhode Island and eastern Connecticut. "In the early '60s, Rhode Island had about the same cancer rate as other New England states. But in the last 12–13 years [the first Connecticut nuclear plant began operating in 1968, although military nuclear installations have been in the area for many more years], Rhode Island rose to the top. It has the highest cancer rate and the highest heart disease rate (outside of Florida) of any state."[1] As if to give further weight to these conclusions, authorities periodically reduce the maximum allowable radiation doses, both for workers and for the general public. Most recently, in 1971, permissible reactor releases were lowered by 99 percent.[2]

What happens when a dose of 600 r, the amount guaranteed to

be fatal, is received all at once by the whole body, is vividly described by the British Medical Research Council:

The first effect . . . is a sensation of nausea developing suddenly and soon followed by vomiting and sometimes by diarrhea. In some people, these symptoms develop within half an hour of exposure; in others, they may not appear for several hours. Usually, they disappear after two or three days. In a small proportion of cases, however, the symptoms persist; vomiting and diarrhea increase in intensity; exhaustion, fever, and perhaps delirium follow; and death may occur a week or so after exposure.

Those who recover from the phase of sickness and diarrhea may feel fairly well, although examination of the blood will reveal a fall in the number of white cells. Between the second and fourth weeks, however, a new series of ailments, preceded by gradually increasing malaise, will appear in some of those exposed. The first sign of these developments is likely to be partial or complete loss of hair. Then, from about the third week onwards, small hemorrhages will be noticed in the skin and in the mucous membranes of the mouth, which will be associated with a tendency to bruise easily and to bleed from the gums. At the same time, ulcerations will develop in the mouth and throat, and similar ulceration occurring in the bowels will cause a renewal of the diarrhea. Soon the patient will be gravely ill, with complete loss of appetite, loss of weight, and sustained high fever. Feeding by mouth will become impossible, and healing wounds will break down and become infected.

At this stage the number of red cells in the blood is below normal, and this anemia will increase progressively until the fourth or fifth week after exposure. The fall in the number of white blood cells, noted during the first two days after exposure, will have progressed during the intervening symptomless period, and will by now be reaching its full extent. The changes in the blood count seriously impair the ability to combat infection, and evidence from Nagasaki and Hiroshima shows that infections of all kinds were rife among the victims of the bomb. Many of those affected die at this stage and, in those who survive, recovery may be slow and convalescence prolonged; even when recovery appears to be established, death may occur suddenly from an infection which in a healthy person would have only trivial results.[3]

In her book, *Nuclear Madness*, Dr. Helen Caldicott offers another frightening scenerio of the effects of nuclear plant accidents

Soon after a meltdown with release of radioactivity, thousands would die from immediate radiation exposure; more would perish two to three weeks later of acute radiation illness. Food, water, and air would be so grossly contaminated that in five years there would be an epidemic of leukemia, followed fifteen to forty years later by an upsurge in solid cancers. The genetic deformities that might appear in future generations are hard to predict, but they will surely occur.

Such a meltdown could have staggering consequences. The Union of Concerned Scientists recently conducted a two-year study of a hypothetical "expanded nuclear economy" and concluded that before the year 2000 close to fifteen thousand people in the United States may die of minor reactor accidents. Moreover, they estimated that in the same time period there is a one percent chance that a major nuclear accident will occur, killing nearly 100,000 people; most will die of radiation-induced cancers.[4]

Thus far we have been discussing acute effects of large doses of radiation taken on the whole body or on specific organs and tissues, the kind of doses likely to be received by people in the immediate and intermediate vicinity of a serious reactor accident. Suppose it could be vouchsafed that no such calamity would ever happen: Would you then be able to rest easily?

The answer is most emphatically *no*. The merely commonplace activities of the nuclear industry, the mining, milling, and processing of fuel, the day-to-day operations of nuclear electric plants, the reprocessing of fuel, and the transportation, storage, and disposal of waste fission products, are already contaminating our air and water with radioactivity. If the atomic power program is permitted to proliferate in the coming decade, the presence of these poisons in our environment will reach alarming proportions.

Just how much does our level of "background radiation" have to be raised to be considered alarming? The accretion of evidence garnered in the last twenty years or so by a large number of responsible scientists demonstrates that *there is no radiation threshold below which genetic damage, cancer, or shortening of life is impossible.* Any dose, however small, will take some toll of vital cell material, and it may initiate far-reaching, harmful processes.

While there is no argument that radiation causes many forms of cancer, including leukemia ("blood cancer"), there has been debate over the threshold dose. Research findings have been clouded because scientists still haven't recognized precisely what it is that makes a cell cancerous, nor have they been able to sort out the many factors besides radiation that might, over a period of years, cause or nourish the disease.

In spite of this confusion, two indisputable certainties have now emerged. The first is that *cancer can begin when a single cell is altered* in such a way that its normal self-reproductive powers are affected.

Such a cell begins multiplying rapidly, undeterred by influences that customarily inhibit cell growth. The second certainty is that *the nucleus of a cell*, in which are contained the cell's reproductive mechanisms, *can be damaged by a single particle of radiation*. Although Hermann J. Muller, as long ago as 1927, demonstrated the effects of radiation on reproductive processes—for which he won a Nobel Prize—it has more recently been shown by Robert C. Von Borstel of Oak Ridge National Laboratory that just one alpha particle irradiating the nucleus of an insect egg will kill that egg.[5]

The only question remaining is how long it takes before a malignancy manifests itself. The answer is, it may take generations of the given cell type. The "turnover" in some skin cells, for instance, is about four months, but it may take years or even decades before the "descendants" of a radiation-damaged skin cell form cancerous ulcerations.

The direct relations between cancer and infinitesimal amounts of radiation has been particularly well illustrated in leukemia. Studies of Japanese cases show a straight-line relation between dosage and leukemia-induction down to nearly zero. Even as early as 1948, Dr. N. P. Knowlton at Los Alamos had shown that no more than .2 r—one-fifth of one roentgen—per week of gamma radiation was sufficient to depress white cell numbers, and even smaller radiation doses produced detectable abnormalities in the lymphocytes, as demonstrated by Dr. M. Ingraham II in 1952. Though such abnormalities did not produce immediate ill effects, it was thought that they could well be the forerunners of anemia, leukemia, and other serious or fatal blood diseases. Analyses of leukemia incidence in bomb victims, radiologists, and persons irradiated as treatment for various ailments, made in 1957 by E. B. Lewis of the California Institute of Technology, pointed to a possible threshold *lower* than the amount of radiation we will be exposed to as a result of growing radiation in our environment from the normal operations of nuclear plants and related facilities.

One of the most highly regarded experts on environmental cancer is Dr. W. C. Hueper of the National Cancer Institute. In an exhaustive study Dr. Hueper listed the reports of carcinomas and sarcomas attributable to radiation of one kind or another, natural and man-made. His conclusion is that

the sum total of the numerous observations on occupational, medicinal and environmental radiation cancers cited, indicates that civilized and industrial mankind has entered an artificial carcinogenic environment, in which exposures to ionizing radiations of various types and numerous sources will play an increasingly important role in the production of cancers.[6]

The lung is one of the most sensitive, crucial organs in the body. As might be expected, the effects of radiation are especially harmful. According to Dr. Edward Martell, who has researched lung cancer and plutonium exposure extensively, a serious miscalculation of damage may have been made.[7] Martell, corroborated by John Gofman, states that radiation standards are based on the assumption that plutonium particles are soluble and therefore can be flushed out of the body within a few days. In practice, however, most plutonium is insoluble in oxide form and thus it may lodge in the lung for up to two years, emitting deadly alpha particles all the while. When finally dislodged, after having irradiated perhaps millions of lung-tissue cells (alpha particles can penetrate about one gram of lung tissue each), the plutonium particle may find its way to another organ—the heart or bone, for example—where it may remain, causing further damage in these parts, for a decade or more.

According to Martell's research, moreover, the reason why smokers have a far higher risk of cancer than nonsmokers is precisely because of insoluble radiation. Smoking releases polonium 210, an alpha-emitting isotope of far less potency than plutonium 239. The heat of the cigarette fuses the polonium into insoluble particles; the cilia, little hairlike filters in the lung that clean out many harmful particles, are meanwhile damaged by smoking. Because a smoker's lung is thus left with little protection, smokers are more than ten times as vulnerable to cancer from the nuclear cycle, according to Gofman.[8]

Plutonium is of course far more toxic than polonium; the number of estimated cancers that a pound of plutonium could cause ranges from Donald Geesaman's 9 billion to Gofman's 42 billion smokers.[9] This is over the lifespan of the isotope. To keep this in perspective, let's remember that the world population is about 3 billion at present. In 1975 alone, some 12,000 pounds of plutonium were produced, each of which can cause up to 42 billion cancers.[10]

The connection between radiation and cancer has been brought officially to the attention of the American people and its government through The President's Commission on Heart Disease, Cancer and

Stroke. In Volume II of its Report to the President, February 1965, the commission stated: "The prevention of cancer at present involves avoidance or removal of known environmental causes of cancer. This includes (1) avoidance of unnecessary and avoidable exposure to ionizing radiation and excessive exposure to ultra-violet radiation."

The late Rachel Carson's book *Silent Spring* is widely remembered (though, sadly, also widely ignored) as a study of *chemical* poisons in our environment. A closer reading, however, reveals her equally deep concern with radioactivity. In her chapter on environmentally caused cancer she concludes:

> Today we find our world filled with cancer-producing agents. An attack on cancer that is concentrated wholly or even largely on therapeutic measures (even assuming a "cure" could be found) in Dr. Hueper's opinion will fail because it leaves untouched the great reservoirs of carcinogenic agents which would continue to claim new victims faster than the as yet elusive "cure" could allay the disease. . . .
>
> In one important respect the outlook is more encouraging than the situation regarding infectious disease at the turn of the century. The world was then full of disease germs, as today it is full of carcinogens. But man did not put the germs into the environment and his role in spreading them was involuntary. In contrast, man *has* put the vast majority of carcinogens into the environment, and he can, if he wishes, eliminate many of them.[11]

To what extent radiation shortens human life—and in this context we mean simply premature aging rather than curtailment of life as a result of some specific radiation effect like cancer—is difficult to say. Statistical measurement is impossible, both because of the time factor and the innumerable variables involved. But scientists believe that radiation of any amount definitely ages the population so that it dies from all causes at earlier ages than it otherwise would. They have managed to extrapolate to human values observations made of mice and other experimental subjects, a process that, though subject to some error, can still give us a rough idea of what we can expect for the human condition. Robert S. Stone, in a paper entitled "Maximum Permissible Exposure Standards," states:

> It has been shown with certainty insofar as mice are concerned that exposure to daily doses of X rays of slightly greater than 1 r causes a reduction in lifespan. Boche has shown that the lifespan of the rat is definitely shortened by daily exposures of 0.5 r and probably 0.1 r, the exposures starting at

the time of maturity. On the basis of such figures it was felt that for whole-body exposure the permissible dose should not exceed 0.05 r per day. Even this provides a factor of safety of only 2.[12]

Walter R. Guild, assistant professor of biophysics at Yale University, in his essay "Biological Effects of Radiation," also extrapolated results of life-shortening experiments on mice and surmised that intense single doses of 300 to 400 r will shorten human life by between four and nine days per roentgen of exposure. What about lesser doses? "There seems," he concluded, "to be no dose threshold for the life-shortening effect."[13]

Although radiation exposure is no laughing matter for even the most hardy persons, certain groups are hit harder than others, most notably the very young and the very old. Children, and especially fetuses, have a rapid growth rate, and their cells divide extremely quickly. An affected cell can therefore have a far more dangerous effect—and one that will show up in a much shorter time—on those who are still growing. The elderly must be careful for another reason. As we age, our cells lose the ability to replace themselves with healthy new cells; radiation exposure speeds this process.

Ordinarily a gene, the fundamental unit of heredity located in the chromosomes of cells, is stable. It copies itself unerringly generation after generation. Occasionally, however, it undergoes a spontaneous change, presumably chemical, called a mutation. Then a gene that had hitherto been producing, say, blue eyes suddenly begins to produce brown eyes. From then on, all future generations derived from that gene are brown-eyed. Such gross alterations of hereditary characteristics can also result from breakage of chromosomes, the bodies containing genes, and rearrangement of the broken chromosome parts.

Under natural conditions, mutations are exceedingly rare, on the order of one in 100,000 generations. But Muller, in the 1920s, exposing fruit flies to radiation, managed to increase the number of hereditary abnormalities in their descendants. Subsequent studies in a wide range of plants and animals have confirmed Muller's discoveries: In every organism examined, it has been observed that high-energy radiation reaching the chromosomes will produce mutations.

Not even the Atomic Energy Commission denies the hazard of

low-level radiation. Consider what a booklet published in 1966 by the AEC's Division of Technical Information states:

> The direct effect of ionizing radiation on our chromosomes can be serious. . . . Even if the chromosome manages to remain intact, an individual gene along its length may be damaged badly and a mutation may be produced. . . .
>
> No matter how small a dosage of radiation the gonads receive, this will be reflected in a proportionately increased likelihood of mutated sex cells with effects that will show up in succeeding generations. . . . If a sex cell is damaged and if that sex cell is one of the pair that goes into the production of a fertilized ovum, the damaged organism results. There is no margin for correction. There is no unaffected cell that can take over the work of the damaged sex cell once fertilization has taken place. . . .
>
> That is why there is no threshold in the genetic effect of radiation and why there is no safe amount of radiation insofar as genetic effects are concerned. However small the quantity of radiation absorbed, mankind must be prepared to pay the price in a corresponding increase of the genetic load.
>
> Every tiny bit of radiation adds to the number of mutated sex cells being constantly produced. There is no recovery because the sex cells after formation do not work in cooperation, and affected cells are not replaced by those that are unaffected.
>
> This means (judging by the experiments on lower creatures) that what counts, where genetic damage is in question, is not the rate at which radiation is absorbed, but the total sum of radiation. *Every exposure an organism experiences, however small, adds its bit of damage.* [Emphasis added][14]

What are some of the more harmful defects geneticists fear? A few examples are hemophilia, erythroblastosis fetalis (a blood disease of the fetus or newborn child), familial periodic paralysis, nervous and mental diseases, metabolic and allergic disorders, and certain congenital diseases.

A variety of anomalies, many verging on the monstrous, are possible as a result of radioactive damage to the genes. Gigantism, dwarfism, albinism, clubfoot, harelip, cleft palate, Siamese twins, Janus monsters (two faces on a single head and body), phocomelia (rudimentary limbs), sirenomelus (legs fused with no separate feet), hydrocephalus (grotesquely distended head), hermaphroditism (physical bisexuality) may be induced by radioactive bombardment of reproductive cells.

Of course, in the evolutionary process, harmful mutations—and most mutations are decidedly harmful—tend to eliminate the hereditary lines that carry them, because of sterility, disease, and feebleness. But even if this process of natural selection does erase harmful strains, the elimination process itself entails suffering of every imaginable sort, whole heritages of suffering perpetuated from parent to child to grandchild until the last tormented descendant is laid to rest.

That henceforth man must live in constant dread of a major nuclear accident, which will wreak death and harm on a level potentially surpassing Hiroshima and Nagasaki, is unnerving enough, certainly. But we must realize that even if such accidents are averted, the slow, silent saturation of our environment with radioactive poisons raises the odds that we or our heirs will fall victim to any one of the horrors depicted here, and possibly to some unexperienced in human history.

4
Nuclear Roulette

The members of the International Conference on the Peaceful Uses of Atomic Energy, meeting in Geneva in the fall of 1955, found it appropriate to devote four sessions to the subject of reactor safety. One paper in particular gave estimates of the theoretical magnitude of damage resulting from a reactor accident, and out of these conjectures emerged a more formal study of the possible extent of harm and damage should a mishap occur. That study, published by the Atomic Energy Commission in March 1957, was entitled *Theoretical Possibilities and Consequences of Major Accidents in Large Nuclear Plants.*

The study, undertaken by more than forty leading experts in the sciences and engineering specialties — many from Brookhaven National Laboratory, hence the document's familiar name, the Brookhaven Report — attempted to answer five vital questions:

1. How likely is a major reactor accident?
2. If one occurs, what are the chances that radioactive material will be released into the environment?
3. What factors and conditions would affect the distribution of that material over public areas?

4. What levels of exposure or contamination would cause injury to people or damage to property?
5. If releases of fission products should occur, what would be the scale of death and injury and the costs of damage to property?

To understand the answers, the meaning of the term "major reactor accident" should be made clear. Although nuclear reactors use essentially the same fissionable material as atomic bombs, there is nothing in reactor technology comparable to the mechanisms necessary for triggering an atomic explosion. Thus it is technically impossible for a reactor to explode with the force of an atomic warhead.

That fact does not, however, rule out the possibility of an explosion of conventional size. Loss of coolant or failure of various safeguards could cause the melting or vaporization of fuel. The melted or vaporized fuel could react violently with water or air, or it could produce sufficient steam to rupture the reactor container. Under certain conditions, sodium, used in liquid form to cool certain reactors, can react violently with air. And as we've seen in the Fermi accident, melted fuel can recongeal to form an explosive "critical mass" in one type of reactor.

The force of such an explosion could not only destroy the reactor and breach the containment structures housing it, but it could cause failure of secondary or emergency safeguards as well. This combination of failures would make it possible for gaseous or finely pulverized fission products to be released into the atmosphere, where an unfortunate combination of weather conditions could disperse them over surrounding property and population. People and livestock would be killed and injured, and crops and real estate would be rendered temporarily or permanently useless.

In trying to determine what the chances were of such a disaster happening, and what the specific damage would be, the AEC team was faced with an awesome number of factors—variations in design, construction, capacity, location, local conditions. Were we talking about a boiling water reactor of 100,000-kilowatt capacity, located twenty-five miles from a small city, that has a minor accident on a rainy night in a ten-mile wind blowing toward the population center? Or were we talking about a fast breeder reactor of 500,000-kilowatt capacity, located forty miles from a large city, that has a severe breakdown in a hurricane moving away from the population center? And so on.

The combinations were endless, and the experts of the AEC realized that unless they were prepared to issue a report that was all qualifying footnotes, they would have to make some general assumptions. They therefore hypothesized a typical reactor in a typical location breaking down under typical atmospheric conditions. They presupposed a thermal reactor of 100,000- to 200,000-kilowatts capacity — that is, one capable of generating a maximum of 100,000 to 200,000 kilowatts of electricity. This reactor was located near a body of water, probably a river, about thirty miles from a major city of 1,000,000 population.

It was further presupposed that this reactor was nearing the end of its 180-day fuel cycle when the accident occurred. The fuel cycle is the time it takes for radioactive waste products in the reactor core to build up to the point where they interfere with the reactor's efficiency and the fuel load must be replaced. The AEC team estimated that at the end of the 180-day cycle an inventory of 400,000,000 *curies* of poisonous material, measured twenty-four hours after the accident, would exist in the reactor core. For comparison's sake it can be pointed out that the amount of only one of the many radioactive isotopes contained in the reactor at that time, strontium 90, would be equal to that produced in the explosion of a 3.8-megaton bomb — a bomb 190 times more powerful than the one dropped on Hiroshima. And, to illustrate the potency of radioactivity, it might be mentioned that one *trillionth* of a curie of radon (a gas found in uranium mines) per cubic meter of air is ten times higher than the official maximum permissible dose for miners.

Weather conditions and other factors that might influence the rate and pattern of contaminant distribution were then postulated. It might be dry, for instance, or a gentle rain might be falling. The atmospheric stability might be a typical daytime lapse (meaning the temperature decreases with elevation) with a wind speed of twelve miles per hour; or a typical nighttime "inversion" with a wind speed of seven miles per hour up to fifty meters of height and thirty-five miles per hour above that altitude. In an inversion the temperature *rises* with elevation, meaning it is colder at ground level than it is above. Heat naturally rises, but in an inversion gases and particles that would ordinarily rise into the atmosphere are trapped at or near ground level. Inversions are thus a very dangerous form of weather condition in this context, for a cloud of radioactive stuff, instead of soaring high into the sky and dispersing over a large part of the atmosphere, will be held down and spread over sizable portions of land—a smog, in other words, consisting of radioactive rather than conventional pollutants.

The ground rules thus laid out, the team was solicited for estimates of the odds against three types of accident: (1) one which destroyed or seriously damaged the reactor core but released no fission products outside the reactor vessel; (2) one which released fission products outside the vessel but not outside the building; and (3) one which released them outside the building into the general environment. For our purposes only the third type is pertinent.

Not all of the AEC team members were willing to assign numerical values to the probabilities of such an accident, but the odds given by those who *were* willing were so reassuring that the authors of the report could state flatly: "The probability of occurrence of publicly hazardous accidents in nuclear power reactor plants is exceedingly low." The odds ranged from one chance in 100,000 to one in a billion per year for each reactor.

Supposing this "incredible" long shot materialized, the team was asked to estimate the worst consequences that could be expected. The statistics were no less stupefying for being couched in the typically bland language of governmental reports. *As many as 3,400 people might be killed, 43,000 injured, and as much as $7 billion of property damage done. People could be killed at distances up to 15 miles and injured as far away as 45. Land contamination could extend for greater distances; indeed, agricultural restrictions might prevail over an area of 150,000 square miles.*

The significance of these figures may be difficult to comprehend even for those who have experienced the ravages of flood or fire, earthquake or war. Certainly, nothing remotely comparable can be cited by way of industrial disasters. The worst one of modern American times was the Texas City catastrophe of April 16, 1947, when a ship loaded with ammonium nitrate fertilizer exploded, virtually leveling the city, killing 468, and causing an estimated $67 million in damage. Another major accident occurred in October of the following year, when a five-day smog at Donora, Pennsylvania, resulted in the deaths of twenty persons and the illness of nearly six thousand, some 43 percent of its population. Yet, awesome though these figures seem, they don't begin to approach the dimensions of death, disability, and damage rendered by a nuclear smog cast over a territory possibly as large as 5 percent of the continental United States.

The potential for havoc in our hypothetical reactor, then, clearly has no precedent in peacetime commercial enterprise, and indeed it has few parallels in wartime. It is therefore tranquilizing to remember that the whole thing is merely hypothetical; that the odds against catastrophe are at least 100,000 to 1; and that the experts, in the words of Harold S. Vance, AEC's acting chairman at the time the Brookhaven Report was submitted to Congress' Joint Committee on Atomic Energy, had to "stretch possibility far out towards its extreme limits" in calculating the chances of a major accident.

Suppose, however, that it could be demonstrated that the chances of mishap were nowhere near as remote as the Brookhaven experts claimed? Testimony of a number of experts suggests just that.

The Brookhaven team predicted its odds on the increasing difficulty for an accident to progress from rupture of the reactor core to that of the surrounding vessel and then to breaching of the building's containment structure. This approach implies that each of those units functions in virtual independence from the other two. That assumption is not warranted.

The men who design and build reactors have tried to ensure against the kind of interrelationships among components that would create a "house of cards" type of failure. A completely foolproof arrangement of independent systems, however, is exceedingly difficult to construct. We have it on the authority of numerous experts that this notion really has very little substance. Theos J. Thompson, a former chairman of the AEC's Advisory Committee on Reactor Safeguards, and J. G. Beckerley stated in the introduction to Volume I of their book *The Technology of Nuclear Reactor Safety* that chains of dependency among components may be extremely subtle. "In fact," they said, "reactor designers and operators should beware of the label 'independent.' A structure as complex as a reactor and involving as many phenomena is likely to have relatively few completely independent components."[1]

Even granting that it *is* possible to design totally independent components and systems, the possibility that several *independent* failures can occur successively within an appreciable period is not negligible. Thompson and Beckerley state that at least three independent causes were involved in almost every reactor accident they had studied: a design flaw, a human error, and an instrumentation problem. These and other causes may be operative at the same time in independent areas of the reactor, so that breakdown in one area for one reason and breakdown in another area for another reason might occur coincidentally or as an unexpected result of a third breakdown somewhere else. Thompson and Beckerley have in fact pinpointed thirteen different contributing causes in at least three accidents.

Moreover, such successive independent failures can occur within moments, too fast for even automatic equipment to be of any help — so fast, in fact, as to be considered spontaneous. "A typical nuclear run-away accident may start and be over in times appreciably less than a second," stated three highly regarded experts, C. R. McCullough, N. M. Mills, and Edward Teller, in one paper on reactor safety.[2]

A description of some of the problems that beset just one reactor can give a good idea of the numerous and diverse areas in which failure may lurk. Volume I of *Small Nuclear Power Plants,* an AEC publication, describes ten plants, their design, construction, and operating history. One, the Big Rock Nuclear Plant, about three miles northeast of Charlevoix, Michigan, is a relatively small (75,000 kilowatts) reactor ordered by Consumers Power Company in 1959. It generated electricity for the first time late in 1962 but suffered a number of "outages," or interruptions of operation, because of technical problems. Investigations found the following defects in the reactor: (1) vibrations had jostled a number of screws out of their holes, and these had lodged in key moving parts, jamming them; (2) six out of twelve studs holding down an important piece of shielding had failed, and a seventh was severely cracked; (3) control rods were sticking in position; (4) a valve was malfunctioning for no fewer than twelve reasons (and those were just "some of the known causes"); (5) foreign material was lodged in the rods controlling critical chain reactions; and (6) welds on every one of sixteen screws holding two vital components in place were cracked.[3] Although none of these defects caused an accident, it is not difficult to imagine how failure of one component might lead to failure of some or all of the others.

A profile of a real reactor accident will show how independent difficulties produced a cascading effect that almost ended in catastrophe. As mentioned earlier, on January 3, 1961, three men working inside the containment building of the SL-I, an experimental reactor near Idaho Falls, Idaho, were killed by a blast characterized as a "divergence."[4] According to a reconstruction of the accident published in *Nuclear Engineering* a few months later, the accident was believed to have happened while the men were reconnecting a control rod, which regulates the reactivity of the uranium fuel in the core, to the rod's drive mechanism. Subsequent investigation disclosed that numerous causes had been at work independently or semi-independently.[5]

One was a design flaw: The control rods were designed in such a way that they had to be attached manually to the upper mechanism. In attempting to raise a rod the necessary few inches excessive force had to be applied to overcome stickiness, so when it finally did become unstuck, it jumped. A second design flaw vested excessive reactivity control in the individual rods, so if only a single one were withdrawn too far, the reactor would "go critical."

Another cause was human error, and on a number of levels. The obvious error was that of the operators in pulling the rod out too far, but the supervisory personnel might also be faulted for proceeding with operation knowing that a number of things weren't working properly.

Inadequate instrumentation was another factor. No instrumentation, except for one floor monitor, was in operation at the time of the accident.

A fourth cause was defective equipment: the sticky control rods, for one thing. For another, aluminum casings on cadmium devices used for increasing the margin for shutting down the reactor safely were found to be poorly suited to the heat and corrosiveness of the reactor.

A fifth cause: poor inspection techniques. During the shutdown prior to the accident, the bottom of the reactor vessel should have been inspected for boron, a material that adds substantially to reactivity in the core. It had not been, however.

Acts of God or plain dumb luck can also be independent causative factors in reactor accidents and could have brought about calamity in the SL-I far worse than the deaths of three workers. According to an article in the London *Daily Telegraph* on January 6, 1961, at the time of the incident there was an opening in the plant's outer shell for the accommodation of a crane. This opening should have been large enough for a high level of radiation to escape in the event of such a mishap. "It is remarkable," an AEC official told the reporter, "that radiation did not escape in sufficient quantity to form a cloud which could have caused a disaster." "I can only say we were amazed!" the relieved AEC official remarked. And had the cloud been released during a squall or atmospheric inversion or some other act of God, that might have been *it* for Idaho Falls and environs.

It should be stressed that not only can *components* be damaged by a violent accident, but *the very safeguards designed to protect those components can be damaged as well.* In the SL-ı accident, emergency cooling systems designed to prevent core meltdown, and the emergency system for injecting material that inhibits reactivity, were themselves rendered completely useless by the violent movement of the reactor, which sheared all connecting lines.[6]

Perhaps, theoretically, "redundant" safeguards can be designed and built to plug all loopholes and create truly foolproof independent systems. In reality, however, the complexity and expense of those measures are so high as to threaten defeating the purposes both of people concerned with safety and of people concerned with economics. Representative Craig Hosmer of California, speaking at a congressional hearing, suggested that if designers continue drawing "everything but the kitchen sink on there in order not to have an argument about whether their reactor is safe or not, . . . we are going to accumulate large, unwieldy, and very expensive reactors in which some of the safety features may cancel out twice as many of the other ones and we will have a less safe reactor than we would want to begin with."[7] And in the spring of 1967 an Atomic Industrial Forum task force sent a letter to the AEC expressing similar concern with proliferation of safeguards. *Nucleonics Week* reported the group as taking the position that "carrying engineered safeguards to an extreme is not always in the best interests of safety." The task force went on to bemoan current emphasis on "studying the cause and course of accidents which have only a remote probability of occurrence."[8]

Ironically, then, too many safeguards may be as dangerous as too few. Even if they weren't, they'd still be too costly. The paradox seems to indicate that the public may be damned if the industry does proceed with utmost caution, and damned if it does not. Actually, the whole question of whether reactor components are independent or not becomes irrelevant when matters are examined from another viewpoint.

Suppose we wanted to calculate the odds against a convict escaping from prison. We could first determine how hard it was for him to get out of his cell; then the difficulty of getting out of the cell block; then the difficulty of getting out of the building; then the difficulty of going over the wall. This presumes a sequence of escapes through independent systems, and the odds against a successful series of escapes would be high.

But suppose the prison's boiler exploded, leaving gaping holes in his cell wall, the building wall, and the wall around the prison itself, injuring guards, damaging floodlight and siren systems, and liberating enough prisoners to overcome whatever remaining obstacles confronted them. Under ordinary circumstances each wall in the escape sequence is independent of the other walls. But *all* of the walls are dependent on the integrity of the boiler. The odds against our convict breaching all walls may be 100,000 to 1, but the boiler blast renders the odds meaningless.

To carry the analogy home, utility engineers might design three "independent" systems in such a way that although they all rely on a single electrical power source, they will automatically switch to their own individual generators should that main power source be knocked out of commission by a storm. It sounds good on paper, but could not a flood—brought on, perhaps, by the very storm that knocked out the main power source—swamp the reactor and disable all three auxiliary generators?

The Brookhaven Report pays scant attention to such "incredible" possibilities, yet examples of disasters that successfully bucked all odds come readily to everyone's mind: the sinking of the "unsinkable" *Titanic,* the burning of the "fireproof" Iroquois Theater, the collision of two big airliners over the Grand Canyon. In each instance the odds against the event were fantastically high, and subsequent investigations revealed chains of circumstances so complex and unlikely that one would dismiss them as contrived if they appeared in a work of fiction.

The utility industry itself can point to a prime example: the power failure of November 9, 1965, which blacked out a major portion of the United States' eastern regions. Here was a chain of electric utilities designed to establish independent operaton should failure occur in any sector—yet, through a chain of events so incredible that the odds defy calculation, the lights went out anyway.

We know that nature is no respecter of enormously adverse odds. Everyone lives within range of at least one of nature's destructive forces—earthquake, volcano, tidal wave, flood, heat wave, drought, cold wave, hurricane, tornado, electric storm—and every reactor projected for the future will be within range of one as well. But while experience with these phenomena has enabled utility designers to anticipate them in certain areas of the country, and to anticipate their intensity, nature continues to display perverse reluctance to behave in accordance with man's expectations. Scarcely a day passes without a meteorological or geological anomaly: a hottest day here, a coldest month there, a severest hurricane here, an inordinate number of tornadoes there, lakes drying up for the first time in human memory, snow falling where snow has never before been seen, age-old faults in the earth's crust collapsing inexplicably under seismic strain, long-dormant volcanoes relenting without notice to the pressure of magma seething beneath them. Each of these natural disasters can have fatal effects on nuclear reactors.

Consider tornadoes. They are capable of incredible violence; their rotary winds, achieving speeds of three hundred miles per hour or better, can make a deadly missile out of a piece of straw, uproot trees, and hurl automobiles effortlessly. The partial vacuum created inside its deadly funnel can make buildings explode, sucking their contents into its vortex and raining debris around the countryside. In 1931 a twister in Minnesota carried an eighty-three-ton railroad coach and its 117 passengers eighty feet through the air.

How well will tomorrow's reactors be constructed to withstand tornadoes? In the wake of disastrous damage done to utilities by twisters in April 1965, Philip Sporn, one of the electric industry's leading engineer-executives (now retired), was dumbfounded. Though known for his farsightedness in all phases of system design, he had to admit that "nothing within a tenth of this severity ever happened before." "The devastation happened because we didn't dream it could ever be that bad," he explained.[9]

The following November—one of the lowest tornado-frequency months of the year—a tornado felled a number of transmission lines a half mile from the Dresden power plant in Illinois. The Dresden is nuclear-powered, and it was probably to this plant that Dr. David Okrent, another former chairman of the Advisory Committee on Reactor Safeguards, was referring when he told a congressional hearing: "We do have on record cases where, for example, an applicant, appearing before an atomic safety and licensing board, stated that a mathematical impossibility had occurred; namely, one tornado took out five separate powerlines to a reactor. If one calculated strictly on the basis of probability, and multiplied the probability for one line five times, you get a very small number indeed, but it happened."[10]

Though usually thought of as indigenous to the western and mid-western United States, tornadoes have occurred on the average of twice a year in New York State, six times in Pennsylvania, four in Maine, five in Massachusetts, and once a year each in Connecticut, New Jersey, and Delaware, between 1953 and 1965.[11] Has the possibility of tornado damage in these areas been given serious consideration?

Around the same time that a tornado was shearing the Dresden reactor's lines, operating and maintenance crews in Minnesota were frantically sandbagging the area around the Black Dog power station, threatened by the rising Mississippi River.[12] While a flood is devastating enough to conventional utility plants like Black Dog, necessitating the dismantling and drying out of just about every component that gets wet, the damage to a nuclear station could be incalculably worse, not only in the harm done to the delicate components but in the potential harm resulting from contamination of flood waters with radioactivity. All power plants are going up on one body of water or another; are they designed to withstand floods? Appearing on television after the great power blackout of 1965, Otto Manz, executive vice-president of New York's Con Edison, told of a harrowing experience:

> I mentioned earlier to you gentlemen, . . . I worked in Oak Ridge for Carbide and Carbon Chemicals Corporation and I ran the electrical system down there, and one of the things I was told was don't ever let that plant stop because you might collect a critical mass of uranium hexafluoride in the cascade. . . . And one day the bottom of the Clinch River came up and cut off all our cooling water, and the plant shut down, and that's the only other time my heart was in my mouth 'cause I kept expecting to hear a big atomic explosion. But thank God it never happened.[13]

The Brookhaven Report has become outdated for several reasons. Reactors were much smaller then; a reactor of 1,000 megawatts would never have been contemplated, although this is the standard size for current plants. A megawatt is a thousand kilowatts, or a million watts. The report, as we have seen, considered accidents in plants of the 100,000- to 200,000-kilowatt range—with *one-tenth* the power of a modern plant. This would of course drastically affect the consequences of a worst-case accident. Furthermore, metals used within the reactor are exposed to enormously greater heat and radiation, thus adding stress and increasing the chances of a disaster.

The Brookhaven Report looked at a hypothetical reactor thirty miles from a city of one million. In fact, larger reactors are much closer to population centers. The Indian Point three-plant complex (one of which was permanently closed as a safety risk) is located in the densely populated suburbs of Westchester County, and only twenty-six miles from the more than 7 million people within New York City itself. Imagine the consequences of a full-scale accident there. This is not the only example of poor reactor siting; nearly all are within close range of good-sized cities. The infamous Three Mile Island plants are a little more than ten miles from Harrisburg, the capital of Pennsylvania. Even without these other factors, the property damage estimates because of inflation are wildly underestimated. The dollar will buy only about one-third as much in 1979 as it did in 1957; this alone would raise the potential damage to an awesome $21 billion, without any other changes in WASH-740.

Anxious to ease the fears caused by the Brookhaven Report, the AEC asked for an update in the early 1960s. Unfortunately for atomic promoters, the conclusions were far worse than the original chilling estimates: 45,000 could die, 100,000 could be injured, a damage estimate (in 1965 dollars) went as high as $7 billion, and *an area the size of Pennsylvania could be contaminated* by radiation. Land-use restrictions could last for five hundred years.[14] The AEC kept the report secret, but rumors of its existence had been building. An AEC commissioner had announced that an update was underway,[15] and nuclear opponent Dave Pesonen, who attended the Atomic Industrial Forum meeting where Commissioner John Palfry revealed the existence of the update, wrote to Palfry for a copy of the document.

Palfry's reply enclosed a letter from AEC Chairman Glenn Seaborg to the congressional Joint Committee on Atomic Energy:

> Reactors today are much larger than those in prospect in 1957, their fuel cycles are longer and their fission product inventories are larger. Therefore, assuming the same kind of hypothetical accidents as those in the 1957 study, the theoretically calculated damages would not be less and under some circumstances would be substantially more than the consequences reported in the earlier study.[16]

Pesonen was dissatisfied, and after receiving a letter from Palfry claiming that "no report is in existence or contemplated," he published an exposé in *The Nation*. The AEC continued to disclaim the study's existence.

Finally, on April 18, 1973, David D. Comey sued for release under the Freedom of Information Act. The AEC immediately released documents of more than 2,000 pages, which revealed convincing proof of a deliberate cover-up.[17] Unfortunately, this is the standard policy of the atomic establishment, which will be detailed later.

The Brookhaven WASH-740 report of 1957, and the updated report of 1964–65 (which was suppressed until the early 1970s because of its horrifying conclusions), are by no means the only studies of nuclear power safety. The Reactor Safety Study of 1974, also known as the Rasmussen Report (after its principal author, Dr. Norman Rasmussen), and Richard Webb's *The Accident Hazards of Nuclear Power Plants* (1976) could not be considered reassuring either.

The Reactor Safety Study, or WASH-1400, was hailed by the atomic establishment, despite severe misgivings of the report's many critics. The study appeared to claim that the chances of a full-scale nuclear mishap were less than the chances of being hit by a meteorite. The Nuclear Regulatory Commission swore by the report and used it as justification for the nuclear program from its release in 1974 through early 1979. On January 19, 1979—just weeks before the March 28 accident at Three Mile Island—the NRC repudiated the study, saying that it "does not regard as reliable the Reactor Safety Study's numerical estimate of the overall risk of a reactor accident."[18] The NRC thus vindicated the arguments of Rasmussen's critics.

The most severe and persistent criticism of the Rasmussen study involves its methodology, the "fault-tree analysis" borrowed from the space program. A fault-tree analysis is designed to compare the effectiveness of different systems, to establish relative, not absolute, levels of functioning. An accident is hypothesized, and each factor that could lead to the accident is broken down into its own component causes, continuing until the root cause is reached.[19] In other words, fault trees rely on *linear* progressions. In most nuclear accidents, however, the trouble cannot be broken down to one cause. Rather, a number of independent causes—human error, design flaw, malfunction—combine at once, generally in ways that were not predicted in advance (one example being the formation of a large hydrogen bubble at Three Mile Island #2).

As a consequence, the WASH-1400 calculations negate the major causes of accidents that have already occurred. Amory Lovins, a long-time advocate of small-scale, renewable-energy development, used the Rasmussen techniques to predict one series of events in boiling water reactors—a sequence that had occured at least fifteen times. The Rasmussen techniques predicted that such an incident would happen only once in billions of reactor-years.[20]

A large chunk of Richard Webb's *The Accident Hazards of Nuclear Power Plants* specifically refutes Rasmussen. According to Webb, the study completely neglects the consequences of a power-excursion accident, the industry's euphemism for a nuclear runaway. Webb, who was echoed by Dr. Michio Kaku of City College of New York, feels this would be far more serious than the more commonly discussed loss-of-coolant accident. And it neglects a number of other factors: cascading fuel-rod failures in a power-cooling mismatch accident; steam explosions of nonnuclear materials in the reactor; damage to the reactor from flying pieces of the reactor; realistic estimates of radioactive releases; proper consideration of aging within the reactor, leading to failures in crucial parts; inadequate attention to the degree of human error.

Webb also found numerous other flaws in the report and concluded: "Overall, the Rasmussen Report is unsatisfactory, as many assumptions and assertions are made without specific rationale and supporting references. . . . Is no reliable analysis on which to base a conclusion that reactors are safe, as the nuclear community has done."[21] Interestingly, he noted that "(T)he report contains a notice which disclaims any 'responsibility for the accuracy, completeness, or usefulness of any information' in the report."[22]

Webb's work, based on exhaustive research, computer data, personal extrapolation, and investigation of the few tests that had been done on equipment, notes the terrifying conclusions in the Afterword. He then discusses the many scenarios that could lead to such an event, dwelling only on the actual operation of reactors and essentially ignoring mining, milling, enrichment, transportation, waste storage, and so forth. He notes that reactors are not designed for the worst-case accident. Rather, the "design-basis accidents," upon which reactor safety systems rest, are marked by failure of one component and assume systems such as the automatic shutdown, or *scram*, will work as planned.[23] "In other words," he says, "the reactor plants are *not* designed to avert disaster should a *worst possible accident* occur involving multiple failures."[24] The difference could be substantial. The consequences, for instance, of one fuel-rod failure are much easier to contemplate than the consequences of that one rod causing additional stress and failure of adjacent rods, leading to a "cascading core meltdown" if several other systems should also fail.

According to Webb, we have managed without a *major* accident so far because of our lack of experience. For example, in 1976 we had only about 130 reactor years in plants larger than 400 megawatts; if the thousand reactors once projected had been built and operated for forty years, that would be 40,000 reactor years. (Even Rasmussen's estimates of a one-in-a-million chance has been extrapolated to *one in 33 over a thirty-year period* if based on a thousand operating reactors.[25])

John J. Berger, in *Nuclear Power: The Unviable Option,* mentions several lesser known safety studies. In 1957, the year the original Brookhaven study was released, the University of Michigan, reported the potential consequences of a worst-case accident at Lagoona Beach, site of the Fermi fast-breeder partial meltdown, and estimated that 133,000 people would receive fatal doses of radiation.[26] Berger notes that the AEC tested six small-scale models of the Emergency Core Cooling System, or ECCS. This crucial system is supposed to flood the reactor core in the event of a coolant loss. In these tests, performed in 1970, *all six failed.* In practice, the rupture or leak that removed the original cooling water apparently sucks the emergency water along with it. In a similar occurrence at Browns Ferry #1 and 2 in 1975, reactor operators found many of their safety systems disabled; a meltdown was only averted by jerry-rigging auxiliary pumps, which were never intended for the purpose.

One of the most serious concerns of critics of atomic power is earthquakes, and thereby hangs one of the most incredible tales in the annals of the nuclear industry.

In 1961 Pacific Gas and Electric Company announced plans to build a nuclear power plant fifty miles north of San Francisco on Bodega Head,[27] the first of five reactors on the 225 acres it had purchased for the purpose. A group of about a dozen citizens banded together to oppose it, at first simply on the grounds that land use and conservation would not be well served by the destruction of scenic Bodega Head. PG&E went ahead with its excavations anyway, in anticipation of the AEC construction license.

Because many conservation groups are simply no financial match for powerful utility companies, PG&E didn't anticipate much resistance from the Northern California Association to Preserve Bodega Head and Harbor. But the group wouldn't quit, appearing at County Commission hearings, before the State Utilities Commission, and finally before the State Supreme Court. As it progressed, the movement attracted some well-informed consultants and prominent legal talents, so by the time AEC hearings were scheduled, a formidable party of two thousand indignant, knowledgeable individuals had rolled up its sleeves and dug in for a fight. More important, by that time the original conservation arguments had given way to a far more serious allegation: that Bodega Head #1 was going up smack-dab on top of one of the largest and most unpredictable earthquake zones on earth, the San Andreas Fault.

This famous fault extends from north of Bodega Bay almost to the Mexican border and is part of a larger system of instabilities known as the Circum-Pacific earthquake belt, on which the western coast of South America, the eastern coast of Asia, and the islands of the Pacific, including Japan, are located. Smaller faults parallel to the San Andreas pass through Bodega Head as well. Large earthquakes occurred in the fault area and San Francisco region in 1838, 1857, and 1906, the last being *the* San Francisco earthquake. That quake affected Point Reyes, some fifty miles north of San Francisco and only a few miles from Bodega Head. With each quake the fault has moved between thirteen and twenty-six feet.

The AEC's Reactor Siting Criteria stated explicitly that "no facility should be located closer than one-fourth mile from the surface location of a known active earthquake fault." Yet the reactor was going up 1,000 feet from the edge of the San Andreas, about 320 feet shy of a quarter of a mile.

According to a report by Lindsay Mattison and Richard Daly in the April 1964 issue of *Nuclear Information*, the citizens called in Dr. Pierre Saint-Armand, a professional seismologist from the Naval Ordnance Test Station at China Lake, California. After studying the excavation he reported two crucial findings.

First, the reactor site lay in an area where great strain accumulates in the geologic formations. Here exceptionally high earthquake intensities develop when the material finally breaks, producing the quake movement, or "fling." He even suggested that a great earthquake could be expected within the lifetime of the reactor, a view that has since been confirmed by a number of authorities. In such an event the site would "probably undergo some three or four meters permanent horizontal displacement."

Second, Saint-Armand found that the reactor was being built, in effect, on sand. After describing the ways in which the foundation material of crushed rock and alluvium would shift or transmit shock waves, resulting in serious damage to the reactor, its cooling system and supporting structures, Saint-Armand concluded, "A worse foundation situation would be difficult to envision."

Interestingly, the consultants hired by PG&E itself differed little with Saint-Armand. "It is surprising," the latter said, "in view of the expert advice given by [the utility's advisers] that another site was not chosen and that construction has gone ahead." It becomes less surprising, however, when one realizes that PG&E had already sunk some four million dollars into the morass under the future Bodega Head #1.

Borings taken by the company's consultants showed almost sixty feet of silt, clay, and sand, and one of these experts, Dr. Don Tocher, recommended that if further studies supported this conclusion, "then serious consideration should be given to resiting the reactor in a location where the quartz-diorite bedrock lies at a depth shallow enough that there can be no possibility of wall failure from seismic forces acting on the sands and clay."

PG&E declared it had designed the Bodega Head structure to resist the severest shocks recorded or estimated in California. However, resistance to shock is not the same as resistance to lateral or vertical movement of the ground as a result of slippage in the fault. PG&E stated that its multiple safeguards satisfied the condition that a reactor be designed to withstand the "maximum credible operating accident." Yet in its hazards report no consideration was given to the possibility that the huge shear forces generated in an earthquake could breach *all* containment structures. The company justified this exclusion by claiming that no slippage in that part of the fault had occurred for thousands of years, at least according to their evidence.

The AEC decided to bring in several members of the U.S. Geological Survey, who said (a) the rock in that area could have slipped in recent geologic time, but it left no evidence of the shear; and (b) whatever had happened in the past had no bearing on what could be surmised to happen in the future: "If in some future earthquake surface rupture comparable in severity to that produced in 1906 occurs on the San Andreas Fault, the near surface granitic rock of Bodega Head would be expected to rupture," and "displacement on the order of a few feet, either horizontally or vertically, should be anticipated."

If the AEC was praying for guidance in determining the validity of PG&E's application, its prayers were heeded on Good Friday, March 27, 1964, when a severe earthquake ripped Alaska, killing 114 and casting tidal waves into the Pacific. The Alaskan fault was part of the same system as the one on which Bodega Head #1 was to rest.

Four days later, PG&E filed an amendment to its hazards report that proposed a building structure surrounded by a layer of compressible material, allowing for movements up to two feet without critical damage, "even though the possibility of such movement occurring is not considered to be credible." Seven months later, when the excavation turned up a secondary earthquake fault, the Bodega Head plans were dropped.

On the other hand, the AEC did decide to leave in operation the Humboldt Bay reactor, a 68,500-kilowatt nuclear plant started up in 1963. Humboldt Bay is a little up the coast from San Francisco, but since the reactor doesn't sit *squarely* on any earthquake fault—at least not on any we know of—it has been permitted to go on.

Only *since* the Alaska quake has the AEC promoted the sort of formal geological study necessary to develop conclusive criteria for building reactors on or near sites of known seismological activity. In its *Fundamental Nuclear Energy Research* 1967 the AEC asserted:

> A *prior* knowledge of areas in which geologic-seismologic problems exist *will* facilitate the selection of safe reactor sites. Experience has shown that knowledge of the regional geologic environment is particularly important in areas of high seismicity, such as California, because the earthquake potential of faults and other zones of earth deformation cannot always be determined from investigations in small areas (emphasis added).

This view was strongly reinforced by a report issued at the end of 1968 by the Federal Council for Science and Technology.[28] On December 30 a special group from the council, headed by William T. Pecora, director of the U.S. Geological Survey, stated that only 18 percent of the country had sufficiently detailed geological mapping to give more than the most rudimentary idea of local earthquake hazards. Asserting that experience in this country had given a misleadingly mild picture of earthquake potency, the group recommended a ten-year research program to assess earthquake hazards and set up guidelines for construction policies so that those hazards would be minimized where quakes were likeliest. Referring to the San Francisco earthquake, the report said: "If such an earthquake were to occur in or near a densely populated modern urban area today, the total losses would be measured in many billions of dollars and hundreds, perhaps thousands of lives." The newspaper article describing the council's report does not mention the additional toll that might result from complications following radioactive releases from a quake-damaged reactor.

More recent information bears out these concerns. The Diablo Canyon plant, a two-reactor complex at San Luis Obisbo, California, remains idle although it was completed in 1973. The plant has not been permitted to operate because of earthquake worries.[29] The NRC has begun to pay more attention to the earthquake issue; several plants were ordered shut in early 1979 to retrofit additional earthquake protection.

Seismic upheavals are not just a California concern. The issue is one of many points raised by opponents of a nuclear power plant under construction in Seabrook, New Hampshire, the very plant at which mass citizen opposition in the United States first coalesced into nonviolent direct action. And an article in the WBAI Folio, January, 1979, speculates on the danger to New Yorkers of earthquakes along the Ramapo Fault, which runs dangerously close to the three-reactor complex at Indian Point. Concern was increased when, on January 30, an earthquake on another fault system shook northern New Jersey and parts of New York City.[30]

Such warnings only emphasize the tremulous foundation of seismological knowledge on which tomorrow's reactors are being sited, but other reactor-related sciences rest on equally shaky ground. Indeed, one may read, mingled among the Brookhaven Report's projections of death and destruction, the following admissions:

> The cumulative effect of radiation on physical and chemical properties of materials, after long periods of time, is largely unknown.
>
> Various metals used in reactors such as uranium, aluminum, zirconium, sodium and beryllium, under certain conditions not at present clearly understood, may react explosively with water, also present in many reactors.
>
> Much remains to be learned about the characteristics and behavior of nuclear systems.

The criteria used in establishing [ranges of potential loss resulting from land contamination and evacuation of personnel] are based on meager data.

It seems reasonable to assume that the gaseous elements, halogens and the noble gases, would also escape quantitatively although no determination was made [in a certain experiment]. Unfortunately, such data are not available for the corrosion of more typical reactor fuels such as uranium-zirconium alloy at more realistic temperatures. . . .

It will be clear that the conclusions reached can be little more than educated guesses, since the direct effects on humans of exposures of this character are largely unknown. Similarly, setting definite limits on acceptable contamination levels for land to be used in agriculture is risky because of the incomplete state of present knowledge of the soil-plant-animal-human relationships involved.

Comparatively little is known about the problems involved in living in an environment heavily contaminated by radioactive material.

Exceedingly little is known about the details of atmospheric distribution. . . .

It is important to recognize that the magnitudes of many of the crucial factors in this study are not quantitatively established, either by theoretical and experimental data or adequate experience.

While it can be argued that these disclaimers were written in 1957, a glance at any current report by the NRC on fundamental research will disabuse skeptics of illusions that all the problems of 1957 have been solved, or that many new ones have not cropped up since. Furthermore, it must be remembered that solutions found in the next four or five years will come too late to serve reactors built during this period, or will be inapplicable without extensive dismantling. The utility industry is already complaining volubly about the high cost of "backfitting" new components and safeguards to bring reactors up to date and in line with constantly shifting standards.

One need not be an expert to realize that the whole premise on which the AEC based its findings is essentially meaningless—namely, that generalized conditions or an "average reactor" or a "typical accident" can be formulated. Every reactor has inherent nuclear, chemical, metallurgical, physical, and mechanical characteristics that do not precisely correspond to any other. Each is located at a different distance from a population center, and each population center differs in layout and density from every other. Each stands on a site possessing distinctive hydrological, meteorological, and seismological traits.

Each plant is governed by regulations unique to its own situation, and these are administered and executed by men whose capabilities differ.

The AEC, in short, had to construct the Brookhaven Report on a foundation of assumptions having very little correlation with the infinite variety of specific realities of reactor operations. But while such a statement could be made of any industrial report, such a technique in the evaluation of hazards as great as those of the atomic industry leads to extremely dangerous oversimplifications. The impression one gets from the Brookhaven Report is that a high degree of standardization exists throughout the nuclear utility industry, which in turn suggests strict uniformity of operation, codification of regulations, and intensive quality control. If the AEC can assert that the chances of a major accident in its hypothetical reactor are on the order of 100,000 to 1, the public might complacently be led to believe that the same odds hold true of the reactor being built up the river.

The oversimplification of the Brookhaven Report constitutes the primary fiction from which many myths about safe nuclear power plants have sprung, and the point that cries out for explicit statement in the report is that every reactor constructed tomorrow will be in some sense an experiment—not a stereotype but a prototype. The soundness of individual processes and techniques may have been established in experiments or in other commercial reactors, or presumed from mathematical calculations. How sound they will be in a new combination can be determined only by putting them together in a commercial power plant and trying it out on the public. Such experience is usually gained at the cost of many mistakes, however, and the public must decide whether the magnitude of potential consequences is worth the knowledge acquired. All plants licensed before 1971 were considered experimental by the AEC, which, incidentally, allowed vast research and development subsidies to be funneled into nuclear utilities.[31] The plants built since then have not been standardized; several designs continue to compete, and reactor designers keep attempting major refinements.

That man does not understand many principles and forces in nature is not to his discredit. That he has built empires in the face of his incomplete understanding is to his glory. But that he is gambling with this ignorance and uncertainty, and the fragile yet lethal technology he has woven out of them, against the capriciousness of nature and fate—*this* may very well be to his everlasting sorrow.

5

Nuclear Power: Myths and Realities

In its attempt to make atomic power acceptable to the American public, the atomic establishment has relied on a number of arguments, some of which essentially have no basis in reality. The propaganda machine has been persuasive. It is well funded and bears the stamp of governmental and scientific expertise. This, however, does not prevent the atomic establishment from relying on an elaborate tissue of omissions, distortions, misstatements, and deliberate deceptions, about which nuclear proponents have been startlingly candid.

For example, Frank B. Shants, manager of special projects for the Public Service Company of New Hampshire (owner of the embattled Seabrook twin reactor), writes in the company's *Public Relations Journal:*

> The nuclear debate isn't over whose facts are correct, but, instead, who can come up with the greater hazard and have it successfully perceived so by the people. So forget the facts once in a while. Counter the activists not with facts but with closed factory gates, empty schools, cold and dark homes and sad children.[1]

In this chapter we will examine several of the more persistent myths that have been used to defend nuclear power; others will be examined in more depth in later chapters.

MYTH: No one has ever died as a result of the nuclear program.

A lot of ghosts—and those who *are* dying—would take strong issue with this if given the chance. Directly and indirectly, the nuclear industry *has* been responsible for loss of life far beyond the three men who died horrible deaths in Idaho in 1961.

Dr. Thomas Mancuso, in a study of nuclear workers at the Hanford, Washington, "nuclear reservation," found that nuclear workers are allowed to receive more than twice the amount of radiation that has been shown to double the normal rate of cancer incidence. Mancuso also found that 6 to 7 percent of worker deaths (more than 200 fatalities) over a twenty-eight-year period were directly related to radiation exposure on the job.[2]

In 1963 a dockworker, Edward Gleason, handled a leaking box of radioactive waste—plutonium contaminated chemical residues—and died ten years later of a rare cancer in his hand. Doctors had amputated all the way to the shoulder in an effort to save him. The contaminated box had been unmarked, and Gleason had attempted to stop the leak with his bare hand.[3]

Dr. Ernest Sternglass suggests a connection between routine operation of nuclear plants and increased infant mortality rates.[4] John Gofman and Arthur Tamplin claim up to 32,000 cancer deaths per year would result if every American received the maximum permissible radiation dose.[5]

Karen Silkwood worked in Kerr-McGee's Cimarron, Oklahoma, plutonium fabrication plant. Active in the union (Oil, Chemical and Atomic Workers), she had begun to assemble data on safety violations and other irregularities in mid-1974. By November she had documentation. She had also found her apartment to be contaminated with plutonium. On November 13, 1974, she drove off to meet a high union official and a reporter from the *New York Times*; she never arrived. She was found dead in her wrecked car, and the documents were never found. Although state detectives claimed she fell asleep at the wheel, a private investigator concluded that she had been run off the road.[6] Silkwood's charges of company irregularities were later confirmed, and her survivors recently won the largest court award in history in a civil suit against Kerr-McGee.

Unfortunately, these are not isolated cases. Too many people *have* died, any many more will. Low-level radiation is a slow killer, and we can expect massive increases in cancer and leukemia death rates to begin occurring in the next few years, as the latency periods for those exposed in the early 1960s begin to draw to a close. We can furthermore expect yet another jump around the turn of the century, twenty to thirty years after large numbers of giant nuclear plants began to operate.

MYTH: We need nuclear power to meet our energy needs for the future.

"In 1963 we used half the energy we now consume. . . . We were at least half as civilized," notes Amory Lovins of Friends of the Earth, who succinctly sums up the situation. We not only use far more energy than we need, but we use far more than any other country.

Dave Elliot, an Englishman, points out in *The Politics of Nuclear Power* that we could reduce our energy consumption drastically without affecting our standard of living. Our Gross National Product is about equivalent to that of Sweden, and it is just slightly above those of Germany and Denmark when expressed in dollars per person. Yet, the energy consumption of those countries is only 50 to 60 percent of our own. Furthermore, the amount of power now supplied by atoms is negligible. Although about 13 percent of our electricity is nuclear, the nuclear percentage of the *total energy picture* now stands at about 3 percent, a figure that could be eliminated entirely through conservation alone. Insulating homes and offices would do it. This, moreover, represents a *gross* figure. The nuclear cycle is very energy-intensive; huge amounts of energy must be poured into mining, milling, enrichment, reprocessing, and the actual construction of plants. And because nuclear plants must be located some distance from those who will use the power, much energy is lost in transmission.

Adding in all these factors, the net energy yield from nuclear power generation has so far been a negative number. John J. Berger, in *Nuclear Power: The Unviable Option,* claims that the civilian nuclear program *has consumed five times as much energy as it has produced.* He states: "Cumulative gross electrical output of all U.S. nuclear power plants from 1960 to January 1, 1977, was about 650 billion kilowatt-hours, or 2.1 quads. But the civilian nuclear industry had consumed at least 10.5 quads."[7]

A further irony is in the leveling off of energy consumption. Utilities who contracted for nuclear plants assumed that energy use would continue to rise by an annual increment of 7 percent. However, oil shortages, skyrocketing fuel costs, and a resurgence of interest in decentralized, renewable energy (and conservation efforts) have leveled demand. In 1974, following the Arab oil boycott, we saw the first year in which there was no growth in demand for electric power.[8] Even though many utilities have canceled or deferred new power plants, utility reserves (power available in excess of demand) have been as high as 40 percent during the last half-decade. Clearly, increasng consumption can no longer be used to justify nuclear power.

MYTH: Routine operation of nuclear plants is nonpolluting.

Pollution from normally operating nuclear power plants is harder to detect than fossil-fuel pollution. It is silent, it is invisible, but it is nonetheless real. Two major types of pollution must be dealt with: thermal pollution/water resources depletion and radioactivity. Both heat and radioactive substances are routinely generated *and* routinely released into the environment in nuclear facilities. (These pollutants will be discussed in later chapters.)

MYTH: Nuclear power provides jobs.

Nuclear energy is capital-intensive in the extreme. It costs $105,000 to create one job in public utilities, whereas a job in manufacturing can be created with only $19,500.[9] Thus the capital tied up in *one* job in the nuclear industry prevents *four* people from finding work in manufacturing—of solar energy collectors, for example. And nuclear power is even more capital-intensive than other ways of generating power. During the thirty years a plant would operate, a typical coal plant would require 28,500 worker-years of labor, while a nuke would require only 22,000 worker-years.[10] Many of those jobs apply to building the facility, which means the jobs end when the plant is built. Most of the other jobs are highly specialized and high paying, reserved for those with years of training. So nuclear power does not provide jobs that would have any meaningful impact on a town's unemployed.

Compare this with renewable-resource technologies. The Massachusetts Energy Policy Office estimates that 32,000 jobs in that state alone would be created if $480 million were invested to convert half the state's buildings to solar hot-water heating. This would save 9.5-million megawatt hours of electricity, more than two-thirds the power of the proposed Montague nuclear plant in Massachusetts (which would cost $3.4 billion, or seven times as much, and would produce 13.5-million megawatt hours).[11] Moreover, centralized energy generation robs jobs from many other sectors of the economy. *Electricity has historically been used as a substitute for labor.* The last several decades have seen increased automation in industry and thus rising unemployment. And a study by the Ford Foundation Energy Policy Project predicted that a zero energy growth policy would create *3.3 percent more jobs* by the turn of the century. Interestingly, present energy growth rates are 3.4 percent a year.[12]

MYTH: Nuclear energy is inexpensive.

Rhode Island is the only state in New England (except New Hampshire, which is building the Seabrook plant) without a nuclear power plant. It is no coincidence that Rhode Island has the lowest electricity rates in the region. Even discounting the mammoth governmental subsidies, the atom is a very costly way to generate power.

Those who claim that the atom is cheap generally assume higher than actual capacity factors. A capacity factor is the percentage of the plant's generating capacity that is actually used; if the plant ran at full power for a year, without shutdowns, the capacity factor would be 100. In actual operation, however, nuclear plants have been abysmal failures. Reactors have been shut down frequently for a number of unforeseen reasons, have often been restricted to far less than maximum power output, and have put out significantly less electricity per pound of uranium than had been expected. Moreover, capacity factors decline as the plants grow in size, owing in part to the added stress newer reactors receive from larger doses of heat and radioactivity.

In the early 1970s the AEC report *Nuclear Power Growth 1974–2000* estimated that plants would average 65 percent capacity over the first thirty years of life. It would take thirty-three years for the capacity factor to drop to 39 percent. According to David D. Comey, however, the three oldest commercial plants operating in 1975 had dropped to only 39.2 percent, after only twelve to fifteen years in operation.[13] Comey found an average capacity factor of 57.3 percent, with the average for plants of 1,000 megawatts significantly lower, at 44.5 percent.[14] In other words, *a gross output of 1,000 megawatts would require two plants of 1,000 megawatts each.*

Charles Komanoff of the Council on Economic Priorities found that capacity factors can be expected to diminish 1.5 percent for each additional 100 megawatts of rated power, and to further diminish by one percent a year, after an initial period of debugging.[15] In another study Komanoff found that capital costs for nuclear plants would be 26 percent more (expressed in dollars per kilowatt hour) than that of coal plants, even with scrubbers added to reduce the coal pollution.[16] Even utility industry analyses admit that nuclear power generation is more expensive than coal. *Electrical World* reported in 1975 that nuclear power averaged 18.6 mills per kilowatt hour. (A mill is one-tenth of a cent.) Coal, by comparison, was only 13.6 mills per kilowatt hour, which is 36 percent cheaper![17]

Adding to the nuclear burden is a substantial amount of money that consumers pay into the industry through taxes — real costs that do not show up in the rate base because they are channeled by the government directly into the industry.

Some of these boondoggles are extremely illuminating. Until 1971 nuclear plants were all licensed as experimental units. This meant that they could receive large amounts of federal research and development money — money that might otherwise have been funneled into clean, safe alternatives.[18] Some $5 billion had been thus spent by 1975.[19] (The most alarming subsidy, however, without which no experimental reactor would be operating, is in insurance, which will be discussed later.)

Undercharging by the federal government is also common. In 1976 ERDA, the Energy Research and Development Administration (formerly a part of the AEC, now the Department of Energy) was charging $61.30 to enrich one Separative Work Unit of uranium. By contrast, *Nucleonics Week* estimated that a true cost would be more like $110 per SWU,[20] which would add about 1 mill per kilowatt hour.

The research, insurance, and enrichment subsidies alone have added 25 percent to the actual consumer cost of nuclear power. If such subsidies were included, charges would rise from 4 cents per kilowatt hour to 5.2 cents.[21] The other end of the fuel cycle — fuel reprocessing and interim waste storage — is also subsidized, to the tune of 1.43 to 1.71 mills per kilowatt hour.[22]

This still does not count the costs of decommissioning the nuclear plant. After thirty or forty years of operation, a nuclear plant is highly radioactive, and it becomes a monumental addition to the fearsome nuclear waste problem. According to *Demolition Age* (December 1977) in at least one case the cost of decommissioning was fully equal to the cost of construction.[23] Utilities have yet to set aside any monies for this operation.

The skyrocketing cost of nuclear fuel, which will be discussed later, is the final death knell to any claim of economic viability. And of course the costs of treating health problems resulting from nuclear power, of guarding and storing radioactive waste for thousands of years, and of replacing power unavailable during times when reactors are shut down (often for months at a time) are not figured into this analysis.

In any case, nuclear power is uneconomical in the extreme. By tying up as much as 40 percent of energy research monies,[24] the nuclear program has actually prevented the development of true energy independence. Billions of dollars have been poured down the tubes, and the plants function poorly or sit idle. The nuclear industry has not only consumed far more energy than it has produced, but it has done so at a substantial cost, both through the electricity rate structure and through our taxes.

One wonders why utilities have committed themselves to such dependence on nuclear power generation. The reason lies in federal carrot-and-stick policies. In addition to the subsidies noted above, the government has structured the rate bases of utilities to ensure a firm profit on capital investment. In other words, *a utility will calculate its rates based on a percentage of its capital costs.* A nuclear power plant costs far more to build than any other type of power plant. This means higher rates and higher profits. As if this weren't enough, many utilities have attempted to add nuclear plants into the rate base *while the plants are being built.* They have tried to justify Construction Work In Progress (CWIP) costs precisely because the costs of nuclear power plants are far exceeding estimates, threatening some utilities with financial disaster.

CWIP does not sit well with consumers. Missouri, the first state to pass an antinuclear referendum, in 1976, rejected a measure to include CWIP in the rate base. Similar measures have passed in Hawaii and Oregon,[25] while New Hampshire voters, in 1978, defeated Governor Meldrim Thompson, a rabid nuclear proponent, largely because of his veto of a measure banning CWIP. Joseph Bowring, in *No Nukes: Everyone's Guide to Nuclear Power,* succinctly sums up CWIP:

Right now, without CWIP, consumers pay rates which yield a profit only on assets actually being used to generate electricity. With CWIP, consumers are forced to contribute funds which utilities can't raise in other ways to build plants. Consumers are forced to become investors. They are forced to guarantee that a rate of profit will be earned on the plant even before it is built.[26]

MYTH: The nuclear power program is subject to a consistently high standard of quality assurance at every step.

It should be clear by now that safe, successful reactors require that the whole level of industrial workmanship, engineering, inspection, and quality control be raised well above conventional levels. Even if safety were not a factor, common business sense recognizes that the plant that fails does not earn.

Whatever common sense recognizes, however, industrial management has fallen far short of the standard demanded of men, material, and machinery by this new phenomenon. Routine manufacturing and engineering practices prevail despite the awareness, and the experience, that such practices are utterly inadequate. The absence of an effective regulatory system, furthermore, has permitted those practices to prevail, and continuing disintegration of regulatory policy portends grave consequences if atomic power carries out its blueprint for the coming decade.

The nuclear power industry is a business, and the business of business is profit. While infinitely more sophisticated than the garment, auto, or meat-packing industries, the atomic industry is no more exempt than they from the pressures that force compromise for the sake of satisfying stockholders and customers. That such lapses, where such deadly forces are concerned, can be the forerunners of catastrophe seems to have registered with shockingly little force on government and industry, perhaps because no single individual is able to step far enough away to see how his own unimportant-seeming compromise may be the critical link in a chain of disaster.

In the electrical utility industry, the pressure begins with the consumer; electricity needs must be satisfied. The consumer's pressure is directed at the suppliers of electricity, the utilities, who must keep up with demand. If they fail, they pay three different penalties: (1) public criticism for betraying their commitment to consumer service; (2) stockholder action for losing money, or failing to take advantage of

profitable opportunities; and (3) government intervention, because un-
filled electricity demand leads to overloaded circuits and large-scale
power failures. The choice of power plant type to fill the demand—
coal-fired, hydroelectric, nuclear—depends on many factors, but the
overriding one is invariably economic: How can power be provided
cheaply enough to satisfy consumers, yet profitably enough to satisfy
stockholders?

Whichever form they choose, the utilities must make many steep
commitments and investments: Construction and operating licenses
must be obtained, land and right of way for plant and transmission
lines secured, the community's power network prepared for the forthcoming delivery of electricity, designers and builders engaged, material and components ordered, and so forth. All such commitments
revolve around a deadline, forcing utilities to exact deadline commitments in turn from local, state, or federal licensing agencies, designers, manufacturers, and subcontractors.

In these ways, pressures originating with the consumer radiate
through government and industry. Even from this thumbnail sketch it
should be obvious that those pressures are incredibly complex and
heavy. The opportunities for cutting corners are almost infinite, and
when the pressure gets intense, resistance to temptation may drop.
This is not to suggest that anyone in the nuclear power industry would
set out to commit fraud or deliberately produce inferior work—
though occasional instances of criminal or grossly negligent behavior
in the utility field are by no means unknown. But it *is* to suggest that
all too often, considerations of quality, service, and safety succumb
to seemingly more urgent ones of delivery dates and profit margins.
Even highly responsible executives, and companies with proud records
of reliability, can give way under the frequently merciless pressures
exerted in the business world.

In addition to those strains that the nuclear power industry
shares with other industries, it is subject to some uniquely its own,
arising out of our government's avowed policy to make nuclear-generated electricity competitive with that produced by coal. We have
portrayed how eagerly the government has pursued this goal, and how
heavily it has pressed industry to achieve it. The trouble is that *under
existing technology, a truly competitive atomic power industry is impossible—unless safety margins are reduced.*

By extending the analogy between reactors and prisons in chapter 4, the truth of these assertions can be seen more easily. An architect confronted by a state agency with orders for a prison that is at once escape-proof *and* economical might discover that these values were in conflict; add to that a short deadline, with financial penalties if construction did not meet timetables, and the pressures for corner-cutting might prove irresistible. By substituting thin bars for thick ones on cell windows and doors, brick for granite on walls, low-grade steel for better-grade on exit doors, he can meet the state's deadline with only — to him — minor changes in the specifications. It may cross his mind that such compromises may one day make it easier for a prisoner to escape, but that abstract consideration flees before the state's unrelenting insistence that the building be occupied on the date agreed on.

The architect may not be the only person who compromises: A subcontractor may forge cell bars out of lower-grade metal than was called for in the specifications, or a state inspector may carelessly overlook construction flaws. Obviously, the odds against escape, which probably look high on paper, have been lowered immeasurably by compromises in design, poor material, unsatisfactory workmanship, and inadequate inspection. It can be imagined that with a few more ingredients — an incompetent guard at the cell block door, or failure of the electrical system activating the building's alarm system — a determined prisoner just might make it to freedom. Knowing these things, would you, living a quarter of a mile away from this prison, rest easily at night?

The same thought might have gnawed on the mind of an observer had he attended proceedings held on March 15, 1967, in Room AE-ı in the Capitol Building in Washington, D.C.[27] Here, a little after 10 A.M., the Joint Committee on Atomic Energy convened, the Hon. Chet Holifield, Representative from California, presiding. Testifying before the seven members present was Milton Shaw, director of the AEC's Division of Reactor Development and Technology. The committee had been in session since January 25, conducting hearings on the AEC's authorization bill for the fiscal year 1968.

Shaw had just brought out that EBR-2, the experimental breeder reactor on which, since the Fermi plant failure, all of the government's breeder hopes now rested, had recently suffered an unexpected setback. On March 2, 1967, copper deposits were found in the primary sodium system. This is how Shaw explained it:

MR. SHAW: We think we have found that the prime source of the copper is from the electrodes of an electromagnetic pump. It appears a total of perhaps 11 or 12 pounds of copper has been introduced into the sodium system. We must take steps to prevent further introduction of copper into the system. This, again, may be a problem with quality assurance. I understand that these electrodes should have been plated with a material that is not corrosive in the sodium environment. Nevertheless, they weren't plated properly. We are taking them out now and we will probably either plate them and put them back, or replace them.

REP. HOLIFIELD: Was that a fault of design? Knowing the tremendously corrosive properties of sodium, it seems to me that should have been done as a matter of routine.

MR. SHAW: Mr. Holifield, I think this is probably the most serious type of problem we face in the whole reactor business. There is so much information we have, so much we know that is not being applied, it is heartbreaking.

REP. HOLIFIELD: This was built by the Government, was it not?

MR. SHAW: Yes, sir, but this is both a Government and industry problem.

REP. HOLIFIELD: We had access to this knowledge. Why didn't the engineers in charge of the actual fabrication of this EBR-2 apply the things we know? We can understand why short-cuts might be made in the industry from the standpoint of saving money or from the standpoint of less background knowledge than we have in our in-house capability.

MR. SHAW: Mr. Holifield, I think the environment that has existed too frequently places emphasis on "We must rush through because we are going to miss something if we don't do this job quickly." We keep thinking we can get by whenever a problem develops without correcting the deficiency. Too frequently we find that something wasn't done, such as not taking the radiographs [X ray or similar photographs of components for the purpose of detecting faults], or the radiographs that were taken aren't clear. And the decision is made to move ahead, anyway.

Now these kinds of decisions are made on a very low level, in a sense, even though management may be aware of it, or management may not be. It is very clear that too many problems we are encountering across-the-board are due to not having written procedures or to people failing to follow written procedures.

It can be classified as a difference between a disciplined approach and an undisciplined approach to getting complex jobs accomplished. I believe we have paid a tremendous price for what I classify as an undisciplined approach in our reactor business. . . . I think this point is very important, Mr. Holifield. We must do a better job to exert this kind of discipline in our AEC projects, whether they are built by laboratory or industrial groups.

Shaw also described alarmingly hazardous inspection procedures. Illustrating these with the case of another experimental reactor, the Advanced Test Reactor, even the AEC representative spoke with something approaching indignation:

> The ATR situation is deplorable. I say deplorable because the purpose of the ATR is to provide a test reactor for fuel. We can't even get the parts of the plant together long enough to get it tested properly. Of course, we had a strike out there which did hurt some. But the fact remains that too many of the components weren't built right in the first place. Because of the strike—we took the time to reexamine a number of the components to attempt to correct as many problems as we could during the strike period. We found that the people who apparently inspected and certified suitability of many of the components must have been blind.

Apparently, the inspectors were employees of the companies that made the components and were not obliged to report to the AEC; the contractor, in other words, was free to inspect and pass on his own item without responsibility to the regulatory authority. Asked by Representative Holifield whether the AEC had any system for inspecting items after they are fabricated and certified by industrial contractors, Shaw had to admit, "We do not have such a system set up in our reactor program right now. . . ."

Apparently, quality control and inspection is often a matter of "understandings" and "assumptions," rather than stringent guidelines and forceful regulation. As a result, Shaw stated: "We get into the argument as to 'who is responsible for the malfunction of the valve,' or 'why didn't we tell the designer that the valve had to withstand vibration?' 'Show me the piece of paper that said the valve had to withstand vibration.' The piece of paper does not exist in the manufacturer's hands or in his contract," Shaw declared, "even though the understanding of the need clearly existed between ourselves and our laboratory, or between ourselves and our prime contractor."

Having few standards or strict regulatory procedures to heel to, some firms apparently feel free to cut corners. The extent to which this is done in an industry where whole populations depend on *square* corners is profoundly distressing, as Shaw's testimony demonstrates:

> Let us remember that the reactor manufacturer has the option to change material and quality assurance procedures any time he wants to on the nuclear plant he is proposing to build for the utilities. His incentive may well be economic. Some of the problems are introduced into our develop-

ment programs because of this flexibility. The lead-times associated with many of these safety programs may be 4, 5, or 6 years. Thus as we proceed down the road to develop the research and development information and get an understanding of the failure mechanism, the reactor manufacturers may decide not to use this [improved] material, or, worse yet, even though the reactor manufacturers decided to use the material, we know of cases where the subcontractor making the pipe decided that the specified material was too expensive, or not readily available, and substituted some other material. This is how the problems and the controversies start developing.

As well, Shaw might have added, as the calamities.

Inserted in the record of these hearings was a summary statement of the AEC's participation in the development of criteria, standards, and codes. This document is most disturbing in its revelation of the scant progress made in formulation of these vital standards. Reading it, one gets the sinking feeling that the reactors built to date do not reflect the application of many clear-cut guidelines, and those being built will not reflect many more. In one key passage, for example, the commission tells us that the Nuclear Standards Board of the newly established U.S.A. Standards Institute has taken a number of steps to re-activate the nuclear standards program of the technical societies and trade associations. But "although these actions are expected to yield much increased emphasis to the development of commercial nuclear standards in the future," says the report, "the efforts are just getting underway."[28]

To illustrate the magnitude of the standards problem, and the paucity of progress in that area, the AEC statement declared that *out of 2,800 to 5,000 standards necessary for a typical reactor power plant* in the areas of materials and testing, design, electrical gear and instrumentation, plant equipment and processes, *only about 100 recognized reactor standards had been approved as of March 1967!*[29]

The transcript of these hearings is replete with statements by Shaw of the most alarming nature.

. . . we know a safe and reliable fuel can be designed for the water reactor plants, if one doesn't extrapolate too far. However, there is no question but that the fuel warranties being provided right now far exceed the meaningful information which is available to us through the research and development programs.

Abundant concrete examples show that because of past failures to identify and invoke minimum engineering standards, neither the results of research and development programs, nor design and construction capabilities are being utilized as effectively as they should be.

The price must be paid—sooner or later. Example after example can be cited where failures of equipment have had to be corrected at great cost and time delays after a plant is built and operating.

It is of deep concern when basic and important research and development objectives of the Commission's and industry's programs are not being realized in a predictable manner after costly investments because of insufficient engineering attention.[30]

It is important to note that the Atomic Energy Commission made it clear years ago that ultimate safety of nuclear power plants rests with reactor manufacturers and utilities operating such plants. ". . . We must reemphasize that the primary responsibility for safety must rest with the industry," said AEC Commissioner James T. Ramey before the Joint Committee in 1967.[31] And the Advisory Committee on Reactor Safeguards has stated: "The ACRS believes that it is the reactor owner and operator who bears the ultimate responsibility for public safety and that he should satisfy himself fully in this respect."[32]

The position of the Nuclear Regulatory Commission (and more importantly the federal government since—as this is being written—the President's commission investigating the Three Mile Island accident has recommended abolishing the NRC) on who exactly is to be responsible for public safety is an important question to which every citizen deserves an answer. In this connection, Admiral Hyman Rickover, director of the Navy's nuclear operations, put his finger on one of the tragic weaknesses of our nuclear program when he stated:

"Unless you can point your finger at the man who was responsible when something goes wrong, then you have never had anyone really responsible."[33]

One of the greatest myths of the nuclear power program in the U.S. thus far has been that there has been "responsibility". And Three Mile Island may be just the first major accident to reveal the crucial nature of that particular myth.

MYTH: Nuclear power plants can overcome all deficiencies in quality control and other aspects of safety and be "made safe."

This is only true in the *theoretical* sense that anything could be constructed and operated safely *if all conditions necessary to safety were maintained on 100 percent level.*

The sad fact is that nothing in the real world ever has these perfectly safe conditions. The first page of the October 5, 1979, issue of the *St. Petersburg Times* in which the *Times* editorial asserted that "nuclear plants can be constructed and operated so that they are safe" cited the recall of toy phone cords, spear guns, grill valves, and lawnmowers—items which are all far less complex to manufacture than nuclear reactors. Yet these—and numerous other products including automobiles—are frequently recalled due to errors of design or manufacture.

Certainly as the TMI accident proved, and as the report by NRC inspectors at other nuclear plants cited in "Lessons from Three Mile Island" reveal, we have always been and are still far from the ideal conditions necessary to safety. It is clear that the Union of Concerned Scientists was not exaggerating when it testified:

> ". . . Poor, indeed shoddy, maintenance is common; operator mistakes, including carelessness, frequently occur; crude and avoidable equipment design or installation errors abound. All these bulk large in the accident reports. *One lesson is that in assessing the probability of having a major accident, one must not view things as they might be but rather as they are.*"

Nuclear proponents have, til now, persisted in viewing things as they might be. But though they can continue to ignore all the facts, experience, and evidence until it is too late, neither they nor anyone else will be able to ignore the consequences of seeing only what they wish to see.

6
Human Frailty and Inhuman Technology

Data processing people use an expressive acronym to describe how computer performance depends on human performance: GIGO. It stands for Garbage In, Garbage Out. It means that the quality of a computer's output is only as good as the instructions programmed into it. Stated another way, the flaws in a computer program will reflect exactly those of the men who fashion it.

The same holds true of any other technology: There is virtually no mechanical failure that cannot be traced to a human one. Furthermore, the more sophisticated the technology, the more precise will be the correspondence between the subtlest gradations of wisdom or ignorance, care or negligence, dedication or apathy, and that technology's success or failure. When meters, grams, and seconds are no longer good enough, when the specifications call for millimeters, milligrams, and milliseconds, the demands made on humans are accordingly refined. Minute human lapses that might be tolerable in a conventional industrial procedure will wreck the more exacting one.

And when, as in the case of atomic power plants, the technology is not only exacting but hazardous in the extreme, then a trivial misjudgment or a moment's inattention can spell doom. The word GIGO, then, takes on a special meaning in the context of nuclear power, for the smallest trace of human "garbage" that slips into a reactor's design, construction, or operation may very well manifest itself in a catastrophic eruption of radioactive garbage. We have seen how we are failing to engineer components to meet the cruel tolerances imposed by nuclear reactor technology; now let's examine how well *men* are engineered to meet them.

The sudden surge in orders for nuclear-powered utilities in 1966 caught science and industry by surprise and revealed labor shortages in every sector. An article in the January 1967 issue of *Nucleonics* described as "cliff-hanging" the ability of manufacturers to fulfill orders by promised delivery dates. An official at General Electric called it "touch and go for 1971," and the designers of Three Mile Island, Babcock & Wilcox, said, "If the industry tries to get too many nuclear plants started up at once, there's going to be a bottleneck in constructors."

The yawning gap between orders and capacity precipitated a vast manhunt for new engineering and scientific talent as well as labor on a lower level. Westinghouse, for example, had about 650 professional and administrative employees on its atomic power staff in 1965, but it forecast a 250 percent increase to 1,625 employees by 1969. Unable to find enough hands in the United States, Westinghouse began seeking them in England, Sweden, and elsewhere in Europe — where it found plenty of competition in that rapidly growing nuclear market. "We've gotten about twenty-five already," Westinghouse was able to boast a year or so later. Will Rowland, vice-president of Babcock & Wilcox's Nuclear Generation Department, said his company would have to train many of its conventional-service engineers for nuclear work.

Such explosive expansion negatively affects industrial personnel in a number of ways. We have already seen how it not only leads to errors and dangerous short cuts in the manufacture and assembly of components, but causes inspection oversights and relaxation of management quality control on the part of both those who sell equipment and those who buy it. We have also seen that the shortage of competent personnel in the Atomic Energy Commission itself had forced lowered vigilance in the very agency charged with regulating the industry.

It is not unreasonable to expect, then, that the pressures produced by rapid expansion of the industry will engender shortages in the personnel who operate and maintain nuclear plants, and in qualified supervisors and training directors. Even if a utility starts out with a sufficiency of such personnel, the high salaries operative in a seller's labor market might lead to competitor raids, leaving the victimized utility shorthanded and forcing it to replace highly skilled and experienced men with relatively inferior ones. Even if that does not happen, the usual dynamics of staff turnover will take their toll of top men. Former Rhode Island Senator and JCAE Chairman John Pastore raised this problem, saying, "Unless there is a drastic change in the present military technical management concepts, whereby competent individuals will be assigned technical responsibility for time periods commensurate with the time required to complete a technical project without adverse effect on their promotion opportunities, the high standards of efficiency and safety required will be most difficult to attain."[1]

A "relatively inferior" operator or supervisor may be an extremely able individual whose only disadvantage is that he has not lived with the reactor from the outset. Under conventional circumstances familiarity with every tiny detail of a machine and its technological environment would not be an indispensable condition of safe and successful plant operation; in nuclear technology, however, a very good "second best" is nowhere near good enough.

In fact, it is dubious that "best" is even good enough. Consider first the operator, the man actually at the reactor's controls. In theory he is a thoroughly educated and trained individual whose NRC license guarantees a high level of technical competence. But why should we safely assume that his training and licensing will not be the product of the same kind of corner-cutting and standard-lowering we have witnessed in other areas of the industry, or that company managers will have the time to train operators rigorously, that tests will not be made easier? There is little in industrial tradition to reassure us, and the Congressional hearings held in 1964 on the mysterious disappearance of the nuclear submarine *Thresher,* suggest that the problems are by no means speculative:

REP. HOLIFIELD: Admiral Rickover, I understand that before the *Thresher* incident at least there was quite a bit of pressure to get you to reduce some of your rather strict requirements in the selection of operators and the training of those operators. What is the status of that situation?

ADMIRAL RICKOVER: The attempts toward "degradation of specifications" on personnel—I will use that simple expression—still go on. . . .

Our real problem is in the submarine staffs where nearly all of the people are non-nuclear people some of whom have a deep resentment against the nuclear navy because it has put them out of business. They are constantly trying to get the personnel degraded. It takes a lot of fighting to keep it going.[2]

But even if such unsettling conditions were not transferred into the commercial nuclear power field, it is nevertheless virtually impossible for even the best-trained, most experienced and alert operator to be prepared for, and capable of handling, every contingency that might confront him. If every commercial reactor is a one-of-a-kind phenomenon, possessing a unique and constantly changing personality, how can anyone be certain that an operator will be able to anticipate and deal correctly with eccentricities displaying themselves for the first time, with failures and combinations of failures so "incredible" that no precept in his training, no precedent in his experiences, tells him how to cope with them? As one survey noted, "The importance of operator error as an accident initiator arises from the intimate relationship which usually exists between the design of the plant, the control system, the operating procedures in use, and the personal attributes of the individual operator and his supervision."[3] Is there any wisdom whatsoever in trusting that the personal attributes of the individual operator will *always* be up to sustaining this intimate relationship? A reactor operator is, after all, not a god. Apart from his special skills, he is subject to the same physical, emotional, and mental strains as most other workers, and his every frailty constitutes one more unknown safety margin in any plant.

Licensing regulation 10 CRF 55 provides that the applicant for an operator's license "must pass a physical examination to show that his condition and general health are not such as to be expected to cause operational errors which might endanger public health and safety." Aside from the possibility that the examining physician could overlook or disregard a potentially serious condition, are there not countless ailments that might arise between mandatory examinations? Might not a head cold or severe indigestion or some other indisposition impair an operator's general efficiency or specific physiological functions, such as eyesight or reflexes?

And what of examinations for psychological health? Is it not possible that an otherwise capable operator may bear some latent mental disorder that fructifies after he has been on the job awhile? Dr. Donald Oken, M.D., associate director of the Psychosomatic and Psychiatric Institute of Michael Reese Hospital in Chicago, points out some rather disturbing observations made in studies of reactor accidents:

> There are ample data indicating that most "accidents" are motivated acts subserving psychological needs. I do not, however, base my concern on data derived from other situations only. A review of reports of past criticality and reactor incidents and discussions held with some of the health personnel in charge reveal a number of striking peculiarities in the behavior of many of those involved—in which they almost literally asked for trouble. This has occurred in the face of an Atomic Energy Commission safety program which has been pushed hard and has generally been highly effective. We know that some individuals have a particular psychological make-up leading to frequent accidents. These accident-prone people sometimes can be identified on psychological grounds. . . . It seems evident, then, that the selection of scientists and personnel for sensitive work should include screening for this factor. . . . Perhaps every installation should have, as a part of its health division, a psychiatrist or psychologist.[4]

One need not get superanalytical to understand that the principle of the death wish, a well-accepted and documented explanation of human behavior, is operative in the accident-prone individual, and might be operative in the psyche of the fellow at the control console. He may normally be well in control of his darker impulses, but suppose he bears one morning a subconscious resentment against the company management for some real or imagined slight, or just wakes up, as everyone does from time to time, simply feeling uninvolved?

Other psychological pressures may also be present, as witness this account of a failure at Puerto Rico's Bonus reactor, reported in the February 1965 issue of *Nucleonics:*

> The operator closed the wrong valve—a bypass valve on a drain trap, in a line that at the time was being utilized as the flow path for steam from the reactor—rather than a valve in a line that was to be warmed up in preparation for the next experiment. Flow of superheated steam was reduced to about 20% of original value for about 2½ minutes until corrective action was taken—after a rapid rise in power level and reactor pressure were noticed. Corrective action was to check the rise in power by inserting one control rod 5 inches. However, Combustion [Combustion Engineering, designer of the reactor] evidently felt in retrospect that this was inadequate,

since one of the nine measures reported to AEC as having been taken to prevent a recurrence is "an effort . . . to dispel an apparent subconscious reluctance on the part of reactor operators and shift supervisors to manually scram [shut down instantly] the reactor."

Aside from the death-wish theory, the "apparent subconscious reluctance" shown by the operators may be attributable to fear that they will be called on the carpet for causing the expensive delays that emergency shutdowns often entail. However one interprets their hesitation, the fact remains that such inaction in a crucial situation could prove fatal.

Even if psychological screening and periodic mental examinations were mandatory in this field, there would still be no way of anticipating temporary lapses of attention or momentary aberrations brought on by financial problems, domestic difficulties, bereavements, health worries, disputes with colleagues or supervisors, hangovers, or just plain daydreaming.

Nor has sufficient consideration been paid to what might be described as the psychopathology of accidents. Dr. Oken stated in the same paper that "accidents tend to occur in clusters during periods in which other signs of psychological stress are evident—the 'accident process' or 'accident syndrome.'" And Theos J. Thompson has noted:

> A surprisingly large fraction of these accidents happened at night, on weekends, or during startups after a vacation period. A typical example is the SL-1 accident which happened on the evening shift January 3, 1961, after a 10-day holiday shutdown. Enough others can be cited to indicate that judgment and alertness may be affected adversely on late shifts or on shifts where morale is likely to be low or where attention is wandering because of holidays or other reasons.[5]

It can be argued that these factors are present in all industries, not just the nuclear utility field. That may be so, but the one difference that makes *all* the difference lies in the magnitude of the consequences. We have pointed out that no industrial disaster can match the havoc produced by a nuclear power plant catastrophe, but this point is of such paramount importance that it is worth the risk of repetitiousness.

Indeed, this unique distinction charges the operator with two special responsibilities: He must not on the one hand become too self-conscious, nor must he on the other become too self-satisfied. We know that excessive conscientiousness can itself bring about the very misfortune it aims to prevent: Might not an operator's acute awareness that a single slipup can endanger thousands of lives and cost billions of dollars of damage prey so intensely on his mind that he freezes in a crisis? Thompson and Beckerley, in *The Technology of Nuclear Reactor Safety*, state:

> Although designers can provide a variety of means for building safety into a reactor, these features may be negated by the operators. A safety device may go out of order, or a gauge yield an incorrect signal, or an operator ignore instrument responses that disclose malfunctions. Unless the operating organization is alert to all signs of abnormality, right down to the most subtle indications, the safety of the reactor is in jeopardy.[6]

What an awful burden to lay on one man's shoulders! Equally portentous is the opposite responsibility of not taking too much for granted. Many accidents occur just when a company is patting itself complacently on the back. Again, the words of Thompson and Beckerley sound a tocsin:

> In reactor facilities one of the chief booby traps exists because there are so many safeties involved. Some accidents have occurred because a relaxed or sloppy crew unknowingly has successively allowed various interlocks and safety measures to be breached one at a time; the logic of the operators always is that there are several and, therefore, the breaching of one is not important. Indeed, the results may be totally inconsequential until that time when the last in a long series is breached, and then the results may be very serious.[7]

It is not always easy to document operator errors in assessing reactor accidents, since those errors may form only a barely perceptible —though no less vital—link in the chain of events. And, naturally, operators are no more eager than anyone else to step forward and accept responsibility for a mishap, especially if it is a costly one. An account of a Canadian reactor accident, however, illustrates how operator misjudgments can compound, if not actually precipitate, an incident of potentially devastating proportions:

> The start-up procedure was to have been normal and the heavy water level was raised to a point for normal operation. An operator in the basement then by error opened three or four bypass valves thereby causing some of the shut-off rods to rise. This movement of the shut-off rods was noticed by

the supervisor at the control desk who then went to the basement and corrected the situation. Apparently, however, some of these rods did not drop all the way back into position, although the signal lights at the control desk were cleared. The supervisor then instructed, by telephone, his assistant at the control desk to push specified control buttons; however, his instructions were partly in error, though normally the error would not have resulted in any difficulty. It became evident from the rate of rise of reactor power that the reactor was super-critical at a time when it was believed to be sub-critical. When the control button to insert the shut-off rods was pushed, it was found that all of the rods did not fall back into the reactor because of the earlier error in instructions, and also because of some mechanical difficulties. When it was observed that the reactor power was still increasing, the valve to release the heavy water into its storage tank was opened and after a few seconds the reactor was shut down.[8]

Some 10,000 curies of long-lived fission products were released into the heavy water as a result of this emergency operation. By comparison, only some 1,500 curies of radium had been produced in the world up to the time of the accident.[9]

Of course, the operator is but one member of a team of employees. Even if his control of the operation is impeccable, plenty of opportunities for error exist for the more conventional breed of worker, and some of these errors could be catastrophic. Although it is imperative that each worker perform his task with perfect thoroughness, caution, and alertness, it is no more realistic to expect consistently high levels of performance from pipe fitters or electricians than it is to expect it from the ablest licensed operator. Because humans are humans, accidents *will* happen. In fact, few of us realize the extent of industrial accidents, but an article in the *New York Times* of September 1, 1968, written by Howard A. Rusk, M.D., reported some startling statistics. In 1966, for instance, 2,200,000 Americans suffered disabling injuries at work, the figures averaging out to 55 persons killed, 8,500 disabled, and 27,000 hurt less seriously *on every working day*. Accidents on the job cause about ten times as many working days away from work as do strikes and other work stoppages. But then that fatal difference rears its head again: The kind of industrial accidents that happen all the time in most other fields, resulting in limited damage or injury or deaths, will have far-reaching effects when they occur in a nuclear plant.

To give some idea of the weight placed on accidents that would be considered trivial in another industry, the AEC was legally obliged to report, and *has* reported in annual summaries, such cases as:

> During the opening of a calorimeter can in an open laboratory, the lid blew off to a height of about two feet above the table top. The sudden release of pressure forced radioactive dust (plutonium oxide) particles out into the room atmosphere. Five employees were in the room at the time. Contamination was dispersed throughout the laboratory and adjacent rooms. Decontamination costs were $4,243. . . .
>
> A chemical process operator received a cut to his finger during cleanup work in a processing area.
>
> A construction force was removing a plastic greenhouse enclosing a hood over a conveyor. . . . The workers, unaware that the tape which they were removing covered old tape which was placed over holes in the conveyor, removed the tape used to seal the greenhouse to the side of the conveyor and inadvertently removed the old tape also, causing the spread of radioactive contamination . . . (cost $5,100).
>
> After an experiment was completed, an employee was instructed to decontaminate glass apparatus used in the experiment. On the way back from the decontamination area, the employee carrying the glass apparatus containing polonium was accidentally bumped by another employee, causing him to drop a small glass tube. The tube shattered, spreading polonium contamination over the immediate area. . . . Operations in the building were down for approximately two-thirds of a day for decontamination.[10]

Minuscule incidents like scratched fingers, lids popping off containers, and people bumping into each other would scarcely be worthy of mention in accident surveys of practically any other industry. That they are officially listed as "serious accidents," that they can shut down a company's operation for almost a day and cost four or five thousand dollars for decontamination, are good indications of how treacherous is the stuff we're dealing with. One would therefore think that no task, however tiny, involving nuclear material would be assigned to a worker not thoroughly familiar with the nature of that material or his duties.

Yet there are instances on record where potentially hazardous jobs were assigned to individuals utterly untrained or otherwise incompetent to perform them. On July 24, 1964, for example, Robert Peabody, an employee at United Nuclear Corporation, was given a plastic container five inches in diameter and sixty inches high to rock on his shoulder to bring about a desired chemical mixture of radioactive material. He was then instructed to pour the contents from the safe shape of the plastic container into a mixing jar eighteen inches in diameter and twenty-six inches high, a shape capable of concentrating the mixture into a critical mass — one in which a runaway chain reaction of explosive energy is initiated. The violent reaction spattered him with radioactive liquid, and two days later he was dead. According to the account, Peabody had no training whatsoever in the field of chemical reprocessing of radioactive material.[11]

If minor accidents committed by conventional laborers are of potentially major significance, what can be said of more serious examples of error and carelessness?

The AEC's annual reports to Congress have revealed some unnerving examples of human negligence in AEC facilities. Perusal of the reports for 1966 and 1967 discloses that 3,844 pounds of uranium hexafluoride were lost because of an error in opening a cylinder; 50,000 pounds of nonradioactive mercury were spilled to the ground in an operational error; 10 gallons of waste solution were spilled on a floor during repair of a valve; approximately one-half pint of plutonium solution was spilled in an elevator; 100,000 pounds of nonradioactive sodium dichromate were spilled from a storage tank and drained into a sewer; a $220,000 fire in a heat exchanger cell of a Hanford Works, Washington, reactor was traced to the accidental tripping of valves by electricians during previous maintenance work, permitting oil to escape from the circulating pumps.[12]

None of these accidents led to disaster; but who will warrant that, with the projected proliferation of power plants and their satellite industries in the coming decade, a similar "goof" will not trigger an event of disastrous magnitude?

Thus far we have been talking about "accidents," mishaps that occur despite the best conscious efforts to prevent them. But not all accidents are "accidental" — some are the product of deliberate human design and planning. Are our nuclear power facilities any better prepared to defend themselves against human hostility or malevolence than they are against human recklessness?

Consider labor trouble. Labor relations have been good, but "good" is not necessarily good enough. In 1967 about 182,000 worker-hours were lost in operations and construction of nuclear-related facilities as a result of strikes against government contractors. The worst of these, accounting for 104,969 worker-hours, was a strike against Douglas United Nuclear by the Hanford Atomic Metal Trades Council, which ran from September 1 through December 12, affecting Hanford's "N" reactor.[13]

These disputes were finally resolved peacefully. But unfortunately, even a nonviolent strike can create extremely hazardous situations at a commercial nuclear power plant. Utilities must work around the clock. A work stoppage in a unionized department or division will, if past experience is any guide, force management personnel to take on unfamiliar jobs, or jobs they haven't tackled for years. The consequences in a nuclear plant are easy to imagine, but the record of strikes against defense installations, hospitals, and public services such as transit, garbage collection, fire and police departments shows plainly that organized labor can display remarkable insensitivity to public welfare if it believes its demands are justified. This is not necessarily a criticism of organized labor, whose grievances against a utility's management might very well be responsible and justified. It is merely a statement of fact that strikes *will* occur, for good reasons or bad, but the best reasons in the world would be of no consequence in the light of a reactor accident at a struck nuclear plant.

Although few labor disputes end in violence, it would take only one ill-considered act of hostility to plunge a territory thousands of square miles in area into nightmare. The act might be perpetrated by one disgruntled employee or ex-employee without the sanction or even knowledge of responsible union leaders. The dispute might take a similar course to one that began in 1966 between the International Brotherhood of Electrical Workers and the Alabama Power Company, a utility powered by conventional fuel. On August 16 of that year negotiations over a new contract with some 2,400 members of the union broke down. Within the next thirteen days saboteurs opened oil valves at four substations in the Jasper area, allowing an essential coolant and insulator to drain from the transformers. By late November of 1966 some forty-five acts of sabotage had been reported by the company, including the shattering of insulators and throwing of chains over high-voltage lines. White-collar workers were assigned double shifts to keep the power flowing, but some found their passage into the plant blocked by strikers.[14]

Had Alabama Power been a nuclear utility, these hostilities would most assuredly have set the stage for a serious accident. Short-circuiting of power lines and destruction of transformers could cut off both regular and emergency power to the reactor and its backup safe-guards. The burdens upon the handful of men operating the plant would enormously increase the possibility of serious error. The management might decide to shut the plant down until the strike was settled, but this does not guarantee that an accident would not occur upon resumption of operations. We have read Theos J. Thompson's observations on the special hazards present in nuclear plants during start-ups; the residual hostility and other emotional excesses affecting returning workers and operators might be contributing, if not causative, factors in an accident occurring at that time.

There is little evidence that commercial nuclear power plant management is allowing for such contingencies in plant plans, safe-guard features, personnel deployment, and other security measures. References not merely against labor sabotage, but against political sabotage, civil disorders, and war are conspicuously absent from the many construction specifications, license applications, and safety reviews examined by the authors of this book. Why?

The reason, incredible as it may seem, is that the Atomic Energy Commission does not require such security provisions as criteria for issuance of construction and operating licenses.

In 1959 the possibility of sabotage in the nuclear field became a reality when—according to Robert Gannon in an article some years later entitled "What Really Happened to the *Thresher?*"—cables on the submarine *Nautilus* were found cut.[15] Nevertheless, the AEC did not require utility applicants to address themselves to the sabotage issue. It is not surprising that when the AEC submitted nuclear plant design criteria to W. B. Cottrell, director of Oak Ridge National Laboratory's Nuclear Safety Information Center, the very first of his long list of general comments was: "The ramifications of civil disobedience, riots, strikes, sabotage, and the like have not even been mentioned. With this vast potential risk in mind, should not the physical security of the plant be considered?"[16]

The AEC did not seem to think so. In fact, not content to leave its position flexible, it issued on February 11, 1967, a notice in the Federal Register of its intention to amend several stipulations in its licensing and review procedures, so that in the future it would be clearly understood that

An applicant for a license to construct and operate a production or utilization facility, or for an amendment to such license, is not required to provide for design features or other measures for protection against the effects of attacks and destructive acts, including sabotage, directed against the facility by an enemy of the United States, whether a foreign government or other person.

By an intriguing coincidence, the AEC's amendment proposal was introduced in the midst of a legal action revolving around this very issue. It happens that on March 22, 1966, Florida Power & Light Company applied to the AEC to build two nuclear power reactors at Turkey Point in Dade County, some twenty-five miles south of downtown Miami. The following August, an attorney, Paul Siegel, wrote the AEC's Director of Regulation discussing the possibility of a bombing attack against the proposed reactors from Cuba. He suggested there might be little difference between dropping a nuclear bomb on one of Dade County's cities and using a smaller conventional bomb to breach the Turkey Point reactor's containment and coolant systems. His letter requested thorough exploration of these possibilities.[17]

On January 25, 1967, the AEC issued a notice that a public hearing on the Turkey Point plants would be held in Miami. It is interesting that in an unrelated but by no means irrelevant event the very next day, a full-time employee of the right-wing Minutemen organization and six other men were arrested by the FBI in Seattle just as they were grouping to carry out a plot that called for, among other things, blowing up a power plant. In any event, Paul Siegel filed for a Petition for Leave to Intervene in the Turkey Point proceedings, asserting that his interests might be affected by accidental releases of radioactivity from the reactors and by intentional attempts to harm them. That the AEC would want to air his contentions seemed reasonable enough to him, considering the AEC's statutory dedication to public health and safety and to the common defense and security. But at the prehearing conference on February 10, the attorney for the commission's regulatory staff said in effect that the AEC had never concerned itself with the effects of sabotage or enemy attack and didn't plan to do so now. The AEC attorney then produced the proposed amendments cited above, which went into the Federal Register the next day. On February 20 the commission issued a Memorandum and Order stating that the Turkey Point hearings would not take up the matters raised by Paul Siegel. On April 27, 1967, provisional construction permits were issued to Florida Power & Light. Siegel filed exceptions. The AEC denied them, justifying its decision with legal and "practical" arguments:

> Neither the Atomic Energy Act nor its legislative history suggests that hostile acts should be considered in the licensing process.
>
> The Commission has the responsibility for interpreting the Act's general standards, and it has not and does not consider the common defense and security and public health and safety standards to refer to enemy attack and sabotage.
>
> The protection of the United States against enemy acts is a responsibility of the defense establishment.
>
> Designing reactors against the full range of the modern arsenal of weapons is not practicable.
>
> The risk of enemy attack is shared by the nation as a whole and Congress has not indicated that reactors are to be treated differently from other structures.
>
> Enemy attacks and sabotage are too speculative to warrant consideration.
>
> Examination of the probability of and protection against attack and sabotage would involve sensitive information.

The fallacies in these arguments are so self-evident that they could be dismissed without comment were it not of the highest importance that every citizen understand unequivocally the consummate danger they represent to life, health, and property. With the help of Paul Siegel's superbly argued briefs, some of these contentions can be examined more closely.

Take the argument that "the protection of the United States against enemy acts is a responsibility of the defense establishment." The Department of Defense's policy, as elaborated in its Directive #5160.54 dated June 26, 1965, is "to develop and promote industrial defense, to encourage industry to protect its facilities from sabotage and other hostile or destructive acts, and to provide industrial management with advice and guidance concerning the application of physical security and preparedness measures." In other words, the Defense Department is not specifically charged with protecting any industry from hostile acts, even an industry in which the explosion of a single conventional-sized bomb in a single installation could have effects as devastating as a multimegaton atomic weapon. To clarify this, part V-C of this Department of Defense policy paper declares: "The protection of property is an inherent responsibility of ownership.

Accordingly, the Department of Defense does not assume primary responsibility for the physical security of privately-owned facilities, federally-owned facilities under the control of other Federal departments and agencies, or facilities owned by any state or political subdivision of any state." In short, neither the AEC nor the Department of Defense were willing to assume the responsibility for nuclear facility security, but both have left it on the shoulders of private owners.

The commission argued that designing reactors "against the full range of the modern arsenal of weapons is simply not practicable." Paul Siegel replies:

> There is a continuum of potential missiles which might be aimed at a reactor containment vessel, going from a spitball delivered by a pea-shooter to a 100-megaton nuclear warhead delivered by intercontinental ballistic missile. It is ridiculous to say that because we might be unable to protect against the ultimate weapon on one end of this continuum, we need not protect against any form of enemy attack or sabotage.

Siegel might also have referred to a passage in the AEC's 1962 Report to the President on Civilian Nuclear Power stating: "Nuclear power could also improve our defense posture; . . . the containment required for safety reasons could, if desired, be achieved at little, if any, extra cost by underground installations, thus 'hardening' the plants against nuclear attack." Why is it no longer "practicable" to build A-bomb-proof plants underground, after the AEC's assertion that this could be done at little or no extra cost? Why has not a single American commercial reactor been designed or built underground? Why has not the AEC steered, if not pushed, commercial utility companies in that direction?

These and dozens of other equally forceful considerations have not prevailed with the AEC. On September 19, 1967, the commission adopted its proposed amendments, making its position with respect to sabotage and destructive acts official. Paul Siegel took his fight to the United States Court of Appeals. The Court upheld the AEC.

In fact, the NRC admitted that *more than a hundred bomb threats* have been received at nuclear plants; explosives were actually found at reactors in two instances. And terrorists actually succeeded in exploding devices at two French reactors.[18] The threat to blow up a reactor with conventional weapons, though, is just the tip of the iceberg of nuclear security issues.

About one percent of radioactive materials has traditionally been unaccounted for.[19] Sometimes losses are far higher: 220 pounds of enriched uranium disappeared from a Pennsylvania fuel fabrication plant, and Oklahoma's Kerr-McGee has lost at least 60 pounds of plutonium.[20] *A workable, if crude, atomic bomb can be made from less than 10 pounds of plutonium.* Recently, *Seven Days* magazine ran an article theorizing that this much plutonium or highly enriched uranium could be made into a bomb contained in an ordinary vacuum cleaner![21]

Even if the atomic industry can keep losses to one percent, one hundred operating plants—fewer than those now operating or under construction—would produce 50,000 pounds of plutonium every year.[22] One percent of this is 500 pounds, enough for fifty bombs.

Obviously a large-scale atomic program will have major repercussions on our civil liberties. While such actions would probably not stop dedicated terrorists—or even prevent the accidental poisoning of a city through incompetence, traffic collisions, or a hundred other factors—the installation of military government, with resulting elimination of freedoms of speech, press, and assembly, would be almost inevitable following a major labor dispute at a nuclear plant, a rise in threats at reactors, and so forth. On civil libertarian grounds alone, then, nuclear power should be absolutely opposed by all those who believe in the Constitution, the Bill of Rights, and freedom of dissent.

The last word on the unreliability of the human element in nuclear power belongs to the renowned Dr. Edward Teller, who stated in 1960:

> A single major mishap in a nuclear reactor could cause extreme damage, not because of the explosive force, but because of the radioactive contamination. . . . So far we have been extremely lucky. . . . But with the spread of industrialization, with the greater number of simians monkeying around with things that they do not completely understand, sooner or later a fool will prove greater than the proof even in a foolproof system.[23]

7

Near to the
Madding Crowd

It might be innocently assumed that in order to build a 1,000,000-kilowatt plant, you simply make everything ten times bigger than a 100,000-kilowatt plant. This is like saying that to build an airplane with a 10,000-mile range, you merely have to scale up ten times an airplane with a 1,000-mile range. In reality, the extrapolations from small systems to large are at the very least enormously complicated, and sometimes they are impossible. A metal capable of withstanding the heat of a small reactor may not hold up under fire in a large one; a valve that contains small-reactor pressures cannot simply be scaled up to accommodate big-reactor ones. The know-how required to build plants 50, 100, 1,000 percent bigger than any built before amounts to an entire new technology. And because experience with 1,000,000-kilowatt reactors has been gained not from test projects sited in remote regions, but from commercial operations only a few miles from population centers, these plants *must* be regarded as nothing more, and nothing less, than gigantic experiments.

Aside from the obvious fact that bigger reactors use more fuel and therefore pose a bigger quantitative threat to the community, bigger reactors also make higher demands on safety features and the time margins in which they must operate to be effective. This was brought out in a statement to the Joint Committee on Atomic Energy submitted in 1967 by Dr. David Okrent, the 1966 chairman of the Advisory Committee on Reactor Safeguards, and Nunzio J. Palladino, the 1967 Advisory Committee on Reactor Safeguards chairman:

> Increases in power density lead to a reduction in the time available for initiation of emergency measures in the unlikely event of a large primary system rupture. The requirements placed on the emergency core cooling systems are more stringent for any size rupture. More precise knowledge is needed both about the course of events during the postulated accident and the damage limits that can be tolerated. This is so because less margin is available with which to handle unexpected events.[1]

Associated with the matter of making fuel cores bigger is that of making them more efficient, for naturally an increase in performance spells an increase in profits. But, as is usually the case, the profits are taken out of the box marked *Safety*. After describing technically how greater efficiency may be extracted from a fuel core, Thompson and Beckerley state: "Pressures to increase core performance tend to force reactor designers to move closer to burnout conditions and to operate on narrower margins of safety as far as fuel is concerned. . . . As the safety margins are narrowed, it may be necessary to devise improved in-core instrumentation or to develop new means of burnout indication. To date efforts in this area have been limited; clearly, as performance has increased, such efforts become more and more necessary."[2]

Even though safety margins have been cut to the bone in quest of higher profits, plant performance has not been up to snuff. As noted in Chapter 5, plants have been operating at only a fraction of their design capacity—and capacity factors decrease radically with increases in plant size. Bigger, in short, is not better. The risks are far greater, but the already limited benefit is further diminished.

Just how critical these issues are may be gathered from the fact that they caused, early in 1967, a member of the AEC's Advisory Committee on Reactor Safeguards to file a public dissent on the Committee's recommendation of a go-ahead on the TVA's application to build two titanic nuclear units at Browns Ferry, near Decatur, Alabama. It was the first time such action had been taken by a committee member in AEC history. The committee, though expressing concern that the average power core densities would be about 40 percent higher than any previously planned light water reactor, that the calculated number of fuel elements reaching undesirably high temperatures was greater, and that the time margin available for putting emergency core cooling systems into operation was less, nevertheless gave its green light to TVA, saying merely that many of the unresolved questions about the plant "warrant careful attention" before an operating license is issued. Stephen H. Hanauer of the Oak Ridge National Laboratory bravely, but futilely, took the position that the committee had not gone far enough. Interestingly, Hanauer's dissent followed by only a few days an indication by Milton Shaw, director of AEC's Division of Reactor Development and Technology, that the atomic power program, while proceeding at great speed, was not always giving proper attention to quality.[3]

These criticisms were found to be dramatically justified a few years later. On March 22, 1975 a serious fire raged at the plant,[4] causing the worst nuclear power accident until Three Mile Island, just four years later. The Browns Ferry accident is worth examining because of the importance of human error. The fire was started when two electricians were probing for air leaks. Although they were working with highly flammable plastics—and although a similar fire had started and been quickly extinguished only two days before[5]—they were using candles for this test!

At later hearings, passing the buck was rampant; those who knew the polyurethane foam was flammable claimed to be unaware of the candles, while those who knew of the candles said they had believed the material to be nonflammable. This comedy of errors, combined with defects in the plant's design, came literally within inches of a meltdown; water level covering one of the two reactors had dropped from the standard 200 inches to only 48 inches before the process was brought under control.

The most important plant defect was in the concentration of important controls in one place. All the electrical equipment governing the operations of both reactors ran underneath the control room—only one for the two plants—on its way to the guts of the reactors. When the fire started in this "cable-spreading room," it not only knocked out most of the routine operational controls for the reactors, but it disabled most safety systems, including the Emergency Core Cooling System, at the same time. Delay in shutting down the reactors turned out to be a severe mistake, for these safety systems had had all of their redundancies wiped out by the same factor. Only by jerry-rigging control-rod pumps and other mechanisms never intended for emergency use was disaster averted.

To further complicate matters, the plant's firefighters refused to follow instructions from the local firefighters: The fire chief of Athens, Alabama, told it like this:

> I was aware that my effort was in support of, and under the direction of, Browns Ferry plant personnel, but I did recommend, after I saw the fire in the cable-spreader room, to put water on it. The Plant Superintendent was not receptive to my ideas.
>
> I informed him this was not an electrical fire and that water could and should be used because the CO_2 and dry chemical were not proper for this type of fire. The problem was to cool the hot wires to prevent recurring combustion. CO_2 and dry chemical were not capable of providing the required cooling. Throughout the afternoon, I continued to recommend the use of water to the Plant Superintendent. He consulted with people over the phone, but apparently was told to continue to use CO_2 and dry chemical. Around 6:00 P.M., I again suggested the use of water. . . . The Plant Superintendent finally agreed and his men put out the fire in about 20 minutes.[6]

Another feature of today's "new, improved" reactor that bears comparison with the Brookhaven Report's modest machine is the fuel cycle. As we have seen, during operation the uranium fuel elements in the reactor core produce an inventory of radioactive fission products that eventually interfere with smooth functioning of the fuel. The fuel elements must then be removed from the reactor and new ones substituted while the old ones are being reprocessed. The time it takes for fuel elements to build up a fission product inventory so high that they must be replaced is known as the fuel cycle. The cycle for the Brookhaven Report's reactor was 180 days—six months.

The replacement of fuel elements in a reactor core is a fantastically delicate operation. Each element is a bundle of slim tubes containing uranium fuel. Once these bundles are irradiated their removal calls for infinite care, for they burn with hundreds of millions of curies of deadly radiation. The replacement process may therefore take a month or longer. At Con Edison's Indian Point #1 plant at Buchanan, New York, it took six weeks to replace 40 of the 120 fuel elements, each of which weighed nearly four tons. Just removing the fifty bolts from the reactor's cover took a week.[7]

Obviously, it is not profitable to shut a plant down for four to six weeks every 180 days, and therefore longer fuel cycles become highly desirable—at least from the accountant's viewpoint. Technologists have thus managed to extend the fuel lifetime of reactors by three or more times the 180 days of the Brookhaven Report's typical reactor.[8] But while this is economically advantageous, it also means that the fission product inventory in a big reactor will be proportionately greater, having had three times as long to build up. Since there is more uranium in there to begin with, the total amount of radioactive poison in a modern nuclear furnace, especially toward the end of its fuel cycle, is staggering. An accident occurring late in that cycle would, if poorly contained, spew far greater amounts of radioisotopes into the environment than one happening at the end of a smaller reactor's cycle. In the case of the accident at Three Mile Island we were very lucky that the reactor was brand new. Although its fuel cycle was designed to be far longer than Brookhaven's hypothetical plant, it had only accumulated a three-month inventory of radioactivity by the time of the accident.

By far the most significant distinction between yesterday's reactor and tomorrow's, however, lies in its distance from population centers. Until the atomic power program began, in the mid-1950s, moving away from the hypothetical and closer to the practical, few gave excessive thought to the matter of how close reactors should be placed to population centers.[9] The main thrust of the government's effort in the field was toward getting reactors built and operating at any cost, just to show potential investors that nuclear power worked. But as the 1950s turned the corner into the 1960s, and the government was able to demonstrate that its small experimental and demonstration reactors worked (more or less), the emphasis began to shift to answering the question: Do they work profitably?

One of the reasons why the government stammered when it tried to answer that one was that it knew that reactors built far from the electricity consumer are uneconomical. Because the costs of purchasing right of way on which to build transmission lines, and the costs of the transmission lines and supporting equipment themselves, are tremendous, long-distance power transmission adds both to the capital and operating expenses of a utility. One authority stated that the costs of rural transmission lines range between $50,000 and $150,000 per mile, exclusive of right of way.[10] So there are powerful economic incentives to move reactors as close to major centers of electricity use as possible, and early in the 1960s industrialists and investors began to reach out for them. In the words of Clifford K. Beck, AEC's deputy director of regulation, speaking before the 1963 annual convention of the Federal Bar Association, "competitive prices for electricity from nuclear reactors began to come within 'smelling range' [about 1962] and efforts to eliminate unnecessary costs, e.g., long transmission lines and large reactor sites, increased."[11]

But there was a slight conflict here, because in its 1962 Report to the President on Civilian Nuclear Power, a major policy statement, the AEC had declared: "For safety reasons, prudence now dictates placing large reactors fairly far away from population centers."

Luckily for the AEC, and later the NRC, its major policy statements on safety have little relevance to actual practice, for while it was preaching caution on the one hand, it was encouraging economic expediency on the other, urging utilities to put up monster-sized reactors and not discouraging them from putting them up near population centers.

In April 1962 the AEC formally adopted and published its guides to reactor site selection. These defined the estimate of upper limits of fission product releases, the containment capabilities of the facility, meteorological characteristics, and potential exposure doses permissible for any site. Of course, "permissible" suggests that these guidelines were enforceable. They were later to prove just suggestions, take 'em or leave 'em.

As Beck pointed out, publication of the guides had a crystallizing effect on siting philosophy. Designers and utility operators, who had been asking what they had to do to get their reactors profitably situated near a population center, now interpreted the guidelines as concrete instructions. "If we can build our containment structure to meet *this* guideline, and design an emergency cooling system to meet *that* one, then our plant will be safe at a distance of X miles from the city," was their reasoning. The AEC's guidelines, which should have kept reactors out, had the opposite effect of attracting them to come in. Beck's exposition of this dynamic is worth repeating in depth:

> . . . a very great change has taken place in the concept of the maximum credible accident. For many older reactors, the maximum credible accidents described were pretty far out in the realms of incredibility, both in the

opinions of the proposers and of the reviewers. The reactors were remotely located anyway, and even if this horrendous incredible accident created no tolerable hazards, then any credible situations, posing lesser hazards, would also be acceptable. Now the situation is different; there is an obvious (economic) reason for the upper limit of anticipated hazard to be as realistic as possible: *assumptions that larger accidents might happen cost money.* (Emphasis added.)

Beck went on to say:

Accident experience to date has been very reassuring. Every reactor, when it is finished and fully evaluated, is confidently expected to operate throughout its lifetime without serious accidents and there is a firm basis for confidence in this expectation. But I would not, and I doubt that many other people would, at the present stage of reactor development, depend on this expectation alone when the reactor is located in the midst of people and the consequences of a misjudgment could be so severe. We are not yet to the point where dependence on the low probability of a serious accident alone can be taken as the only barrier against disaster.[12]

And yet on December 10, 1962—only twenty days after the AEC's report to the President stating "for safety reasons, prudence now dictates placing large reactors fairly far away from population centers," and nine months *before* Clifford Beck stated, "We are not yet to the point where dependence on the low probability of a serious accident alone can be taken as the only barrier against disaster"—Con Edison of New York announced its proposal to build a 1,000-000-kilowatt nuclear plant in Ravenswood, Queens, one of New York City's five boroughs. In other words, a reactor squarely in the heart of a population of millions, a reactor five or ten times the size of our little friend in the Brookhaven Report!

One would think the AEC would have come down on this reckless proposal with hobnail boots, for it violated every scruple the commission had ever enunciated, including the one it had made less than three weeks earlier. Yet for thirteen months the committee walked around it, considering.

Why did the AEC hesitate? Because, as we have seen, atomic energy was on trial in 1962, and the AEC was urgently looking for a power plant that would vindicate all the time, money, and labor this nation had sunk into making nuclear power facilities economically viable. *Nucleonics*, reporting the month-to-month developments, quickly went to the heart of the matter, stating in its January 1963 issue:

> By so doing, Con Ed clearly threw down the gauntlet on the long-controversial question of siting reactors near population centers. If the site, in the middle of the United States' biggest city, is approved and the plant built, Con Ed will have succeeded in making nuclear power an entirely conventional, routinely accepted thing.

And, in its next month's issue:

> The process will be closely watched by the entire nuclear industry. . . . For if nuclear power is to become commonplace and gain an equal footing with coal and oil power, it must gain acceptance for plant location close to or in large cities.

Opposition to the plant was ferocious, and it took on a somewhat sensational aspect when none less than David E. Lilienthal, the first head of the AEC, declared that he "would not dream of living in Queens if a huge nuclear plant were located there."[13] He was promptly attacked for his "head in the sand" attitude and, since he hadn't been associated with the commission for thirteen years, for fuddy-duddyism. One newspaper ran an editorial saying "the generation of nuclear power for civilian community use has passed the visionary and experimental stages and progress in the development of such power should not be barred by merely uninformed emotionalism or antiquated fears."[14] And Con Ed's chairman actually bragged, "We are confident that a nuclear power plant can be built in Long Island City, or in Times Square for that matter, without hazard to our own employees working in the plant or to the community."[15]

Just what processes worked to cause Con Ed to withdraw in January 1964 cannot be clearly ascertained, but what they said in effect when they pulled out was "We'll be back."

Utility arrogance in this area has been curtailed, in part because the NRC now requires reactors to be sited in less populated areas[16] (although this has no effect on plants already sited). Now, there is a new arrogance, shown by this comment: "Rural sites are desirable because of atmospheric dispersion, ease of evacuation, and the *fewer people living in the area means there are fewer people to intervene in public hearings*"[17] [emphasis added].

Con Ed's test was both too sensational and too premature. But while Con Ed was licking its wounds, a number of other companies were more quietly setting up shop close to population centers around the United States, and by 1967 Dr. Beck would have to confess to the Joint Congressional Committee on Atomic Energy that "Connecticut Yankee, Connecticut Power and Light Company, the Malibu location, the Rochester Gas and Electric and Con Edison No. 2 of New York, and others, and Niagara Mohawk have been approved with lower distances than our general guides would have indicated when they were approved."[18]

The implications of reckless siting policies are not difficult to imagine. A breeze just a few miles per hour higher than average, or an unpredicted shift in wind direction, might be all that is necessary to convert a local accident into widespread disaster. For example, though Bodega Head, California, is fifty miles up the coast from San Francisco, some twenty miles farther from a major population center than the distance hypothesized in the Brookhaven Report, an accident at the proposed Bodega Head power plant might have seriously affected San Francisco's population under weather conditions only slightly different from those postulated in the report. The Bodega Head Association one day released fifteen hundred balloons from the reactor site, each carrying the message: "This balloon could represent a radioactive molecule of strontium 90 or iodine 131." Some landed in Marin County, Petaluma and Napa and a few in Richmond, all in the San Francisco Bay area.

It was calculated that external doses of radiation to San Franciscans on the first day of exposure could be almost four times the *yearly* permissible dose, and within ninety days about twenty-five times the yearly permissible dose. Radiation levels of that intensity would, if Brookhaven Report criteria were applied, officially call for "probable destruction of standing crops, restrictions on agriculture for the first year." *That* in some of the most valuable farm country in the United States.

It is obvious that the government has not addressed itself to the philosophical questions relating to nuclear plant siting and safety. Aside from the big question of putting mammoth reactors near great cities, a number of lesser but nonetheless crucial ones remain unexplored. For example, how does one justify placing a reactor near a low-density area, knowing that its population will soon swell to high density? Or, by what logical process does our government conclude that it is "better" to risk destruction of a small rural town of 10,000 people than a large city of 1,000,000? And would not a "minor" release of radioactive material into a highly populated area cause as much harm as a major release occurring in a reactor much farther away from the same city?

But even if we can set aside these moral questions as easily as our government seems to have done, the technological questions remain as insoluble today as they ever were. Testifying before the Joint Committee on Atomic Energy in the spring of 1967, Dr. Clifford K. Beck, AEC's deputy director of regulation, stated:

> The actual experience with reactors in general is still quite limited and *with large reactors of the type now being considered, it is non-existent.* Therefore, because there would be a large number of people close by and because of lack of experience, . . . *it is a matter of judgment and prudence at present to locate reactors where the protection of distance will be present.* [Emphasis added.][19]

Beck's statement is mild compared to that made in the same hearings by Nunzio J. Palladino, chairman of the AEC's Advisory Committee on Reactor Safeguards for 1967, and Dr. David Okrent, the chairman for 1966:

> . . . the ACRS believes that placing large nuclear reactors close to population centers will require considerable further improvements in safety, and that *none of the large power reactors now under construction is considered suitable for location in metropolitan areas.* [Emphasis added.][20]

But these grave, grave doubts on the part of these highly placed officials no longer hold sway with a utility industry now strongly attracted to the smell of profit. The giant reactors are going up everywhere, and they are going up just outside our city limits.

After its defeat over the Ravenswood site, Con Edison was, as threatened, back in action. It proposed to put 4,000,000 megawatts of nuclear power in a four-plant complex on David's Island, just two miles north of the New York City line, and in the heavily populated city of New Rochelle. The plant was announced in 1968, but by early 1972 Con Edison had been forced to cancel its plans; citizen pressure from the Environmental Action Committee of Co-op City (located in a housing project on the New York side of the line, with a population of more than 57,000) and other groups was rapidly mounting, and Con Edison decided to retreat.

Had the populace been less informed, this monstrous complex would have been within walking distance of the most densely populated city in the country, at a site visible from the Empire State Building.

New York City is also a city of universities, and some have experimental reactors. Columbia University has been trying to turn on its Triga Mark II research reactor for years; it has been blocked by city officials and massive protests. The reactor is located on Amsterdam Avenue, in one of the most crowded areas of the city. Moreover, Manhattan College in the Bronx has a similar reactor. And every few months, rumors surface that it is actually operating in the City of New York.

8

No Place to Hide

Considering what has been said so far, no one could be blamed for being seized with the impulse to flee. Unfortunately, even if it *were* possible to relocate out of range of all atomic installations planned for the next few decades—which is doubtful in itself—the pursuing atom will eventually place even the remotest regions in jeopardy. There will be no place left to hide.

Thousands of tons of nuclear fuel are transported by truck, train, ship, and plane to stoke reactor furnaces around the globe; giant reactors propel naval ships across and under the oceans and into ports a stone's throw from the hearts of the world's greatest seaside metropolises; load after load of seething radioactive poison travels through or near our nation's towns and cities on their way to underground storage sites. All of the problems, perils, and probabilities characterizing stationary nuclear facilities exist in mobile ones too. In addition some are unique to mobile nuclear technology.

Massive efforts to power aircraft and land vehicles by means of nuclear reactors have not gotten past the experimental stage, thankfully. For once, recognition of hazards seems to have been a controlling factor. From what we've seen thus far, however, it obviously takes only a little economic incentive to blind the most sensible man to risk, and one suspects that as soon as someone develops a way to make nuclear air and land propulsion economically feasible—or, to put it more accurately, to make it *appear* economically feasible—caution will undoubtedly be thrown to the winds.

There are, however, a number of flying nuclear reactors—satellites in deep space. At least twenty-eight American and nine Russian nuclear satellites have been launched;[1] at least three of these have come flaming back to earth, scattering radioactivity into the biosphere. Most recently, a Soviet nuclear satellite scattered large, intensely radioactive fragments over a mercifully sparse area of northern Canada. Earlier, in 1970, the aborted Apollo 13 moon mission had experienced a similar crash.[2]

An ill-considered and ill-fated experiment occurred on April 21, 1964, when the AEC "lost" 2.2 pounds of plutonium 238, described as a "fiendishly toxic" isotope, when a transit navigational satellite failed to go into orbit. The plutonium's function was to run the satellite's electrical systems, but because someone forgot to throw a switch, the rocket went awry. For some time nobody knew quite where it had gone. Some experts said the rocket had reentered the atmosphere and burned up along with its nuclear payload. But nobody actually saw the rocket reenter, and, the commission acknowledged, "anomalies" can sometimes occur in which metal parts reach the earth without burning up.

The maximum permissible dose of plutonium 238 in the bodies of atomic workers is two billionths of a gram. For all anyone knew, enough of the stuff to reverse the Afro-Asian populaton explosion was mucking about the Eastern Hemisphere. Eventually, unusually strong traces of the element were detected in the upper atmosphere, indicating that the payload had indeed vaporized. Some scientists hailed the discovery as a good thing because it afforded them an extraordinary opportunity to track meteorological conditions. At the same time, humanity's radiation budget, already progressing toward exhaustion, was reduced to the tune of 2.2 pounds of plutonium.[3]

Hopes for the success of these experiments have overshadowed concern regarding the effects of their failure. In one test, a B-36 bomber was loaded with a 1,000-kilowatt reactor. Forty-seven 300-mile flights, between Fort Worth, Texas, and Roswell, New Mexico, were carried out. The reactor was not used to power the plane but just to find out some things about radiation behavior under airborne conditions.[4]

A lot might have been learned about radiation behavior under crash conditions too, but luckily no such thing occurred. Later on, the crashes of nuclear weapons-bearing military aircraft in Spain and Greenland provided ample data to fill our information gap on the behavior of radioactive material—and of humans—when nuclear payloads fall from the sky.

While the equipping of planes and cars with nuclear reactors is still thankfully low on man's agenda for systematic self-destruction, the outfitting of ocean vessels with them is something else again. Here, near-total disregard for the resources of the sea and the welfare of the people on its shores presents a whole new dimension to the issues before us. Although military applications of nuclear material aren't strictly germane to this book, a brief look at naval experience with atomic reactors can shed much light on the dangers of a sea-borne nuclear technology.

The special functions of warships demand special reactor considerations, and these in turn present special problems, problems of containment and cooling, problems of radioactive waste control, and of course problems of shielding against military action.[5] A power reactor operating in a ship does not possess the margin of an exclusion distance, the safe zone between it and the personnel operating it. Therefore, adequate shielding of the containment vessel is a must. Yet excessive shielding adds weight, a distinct disadvantage either in a merchant or a military vessel. Leakage rates of radioactivity must be severely restricted, yet because the structure is subject to movement, there will be difficulty maintaining tightness of joints, pipes, and cables. Ventilation control, especially in a nuclear submarine, is a most important matter. The facts that naval reactors commonly use a more concentrated form of uranium fuel and that the cooling system functions by means of pressurized water present unique challenges to technologists. These facts take on new significance when we consider the potential tally of victims should a serious reactor eruption occur in a large port city. Recognizing this threat, some foreign governments have closed their harbors to nuclear ships or have strongly protested their entry despite the most vigorous reassurances on the part of the Navy and AEC.

These reassurances have been undercut, however, by alleged instances of radiation leaks by our nuclear ships anchored off the shores of host countries. In May 1968, for instance, Japan's Premier Sato told the United States that, because of increased radioactivity measured in the waters of Sasebo Harbor during a visit of the *Swordfish,* a nuclear-powered sub, Japan could no longer permit·American nuclear vessels to call at her ports unless their safety was guaranteed.[6]

That the radioactivity of this alleged discharge may have been low is utterly beside the point. Of far greater pertinence is the proximity of the reactor to a population center. For even if land reactors were located at sensible distances from metropolises — if there *is* a sensible distance — the presence of a nuclear-fueled ship in a harbor represents a flagrant violation of every official siting guideline in existence. *It means atomic reactors within, or virtually within, city limits.*

Japan has had trouble with its own nuclear navy, composed of one ship, the *Mutsu.* Weighing more than 8,000 tons, the *Mutsu* cost $133 million to build. It is the only Japanese reactor constructed without assistance from Great Britain or the United States. Citizen protests delayed the *Mutsu's* sailing from 1972 to August 1974. When the *Mutsu* finally sailed, it immediately ran into trouble. Radioactivity was detected to be leaking from the top of the reactor, possibly because the top reactor shield was inadequate to stop the fast-moving neutrons. A variety of fixes were attempted, including covering the reactor top with socks filled with polyethylene. Nothing worked. But, when the *Mutsu* attempted to return to port, it found its return blocked by citizens laying sandbags at the harbor mouth. The ship was finally allowed to return to port on the condition that a new home port be found within six months and the old facilities be dismantled. At least ten cities refused to be a port for that submarine.[7]

What can happen to a nuclear ship in harbor? A better question is, what *can't* happen to one? First, design and workmanship errors could militate to cause a reactor explosion capable of breaching containment structures, releasing fission products into the water and air. Second, sabotage is a possibility for nuclear ships, and it should be considered a likelihood, even if those ships are nonmilitary. Not only are they strategic military targets, but an incident in a foreign port could have profoundly damaging effects on our political relations with the host country, its neighbors, and all other nations where our fleets cast their anchors. Many demonstrations and bloody protests, some anti-American and others antinuclear, have been aimed at our atomic submarines overseas, and one of these could result in an attempt to storm and wreck one of them dockside. Damage to safeguards could conceivably bring about the very results feared by the demonstrators.

A third possibility is human error. Manpower shortages in the Navy can lead to the same hazards mentioned in connection with land-based reactors. In 1963, according to a *New York Times* article by military expert Hanson W. Baldwin, "many officers aboard nuclear submarines were being forced to serve 'inordinately long' tours of sea duty and many were under great strain, because there was an insufficient number of qualified officers to relieve them."[8] Indeed, the nuclear submarine force had tripled the number of personnel required. The Navy initiated an accelerated training program that year, but in March 1965, 124 naval officers out of 493 men specially selected for nuclear power training resigned from the Navy, or they indicated their intention to resign when their hitches were up. This threatened to plunge the nuclear fleet back into manpower shortages that could one day open the sea valves to a flood of disastrous human error.

It is even possible that hurricanes, typhoons, and earthquake-induced tidal waves could cause unexpected and unexpectedly severe damage to a nuclear vessel in port. By a weird irony, in the very same week that Japanese reporters screamed "Yankee, go home!" at American experts investigating allegations of the *Swordfish* radiation leak, Japan's biggest earthquake since 1923 shook a 600-mile stretch of seaboard, touching off tidal waves and causing widespread damage.[9]

Nuclear vessels can also suffer damage on the high seas, of course. At first glance such damage would appear to threaten only the crew, but a reactor-driven ship partially immobilized by mishap or enemy action would be an extremely hazardous proposition to bring into port, for a nuclear runaway in progress in the ship's bowels could fulminate just as the vessel came within range of population concentrations. If the ship sank, damaged fuel elements would release dangerous radioactive contaminants among shellfish, food fish, and other marine life on which man directly or indirectly depends, to say nothing of subtler genetic effects the radiation would have on marine biology and ecology.

Moreover, it is well documented that radiation concentrates as it progresses up the food chain.[10] The dose received by a human being, then, is perhaps thousands of times stronger than that received by the plankton, which first absorbs it. Each contaminated organism is eaten in large numbers by the next animal on the food chain, and very little of the poison is excreted. The net result is a possibly deadly dose appearing out of seemingly infinitesimal sources. (The same phenomenon has been documented with nonradioactive poisons, such as insecticides, as Rachel Carson pointed out in *Silent Spring*.)

At present, the Navy operates more than a hundred nuclear submarines, along with at least eight surface ships (aircraft carriers and cruisers), many of which have several reactors. Many more are under construction, for both surface and submarine use.[11] The ships in our nuclear fleet are heavily safeguarded and armored, but they are by no means indestructible in this age of superweapons. Even setting aside the question of damages sustained in warfare, however, there is sufficient record of damages sustained in noncombat situations to make potential investors in commercial nuclear ships hesitate—at least for a few moments. On October 13, 1965, for example, two nuclear-powered subs, the *Sargo* and the *Barb*, collided off Oahu, Hawaii, while on submerged maneuvers. There were fortunately no casualties and only superficial damage. Still, if *two* nuclear subs could bump into each other with all that ocean to navigate in . . .

With respect to the merchant fleet, the annual reports issued by the Liverpool Underwriters Association, a leading British insurance group, state that every third merchant ship in service in any year will suffer some sort of disabling event. According to the 1965 edition of its survey, for example, 8,317 out of 24,028 merchant ships of five hundred tons or more throughout the world suffered collisions, weather damage, stranding, fires, explosions, damage to machinery, and contact damage in 1964. This is 34.6 percent of the total. This figure was up 1.4 percent from the previous year and 2.4 percent from 1960. And these statistics don't even include the 117 vessels listed as totally lost that year.[12]

What is the status of our nuclear merchant marine? At present the only American nuclear-propelled merchant ship in existence is the *Savannah*, belonging to First Atomic Ship Transport, Inc., a wholly owned subsidiary of American Export Isbrandtsen Lines. Shippers everywhere had watched the *Savannah*'s performance carefully in the hope that it would fulfill its promise of high efficiency and low cost. But early in 1967 it was announced that the streamlined beauty would be retired: It just wasn't pulling down a profit.[13]

In addition to the technical difficulties of staffing the *Savannah* with highly trained crew members — and note how this problem comes up in sector after sector of the nuclear industry — the company's management had been plagued with heavy liability expenses. A lot of countries had simply forbidden the *Savannah* entry until it was indemnified against serious nuclear mishap. And that meant a guarantee of up to $500 million per country.

On November 18, 1967, it was suggested at the annual meeting of the Society of Naval Architects and Marine Engineers that a nuclear merchant marine could be economically established if safety was deemphasized. George L. West, Jr., professor of marine and nuclear engineering at the University of Michigan, and Lieutenant E. J. Roland of the Coast Guard proposed that "by simply doing away" with many of the present design and operational safety requirements, "many of the first cost items, such as containment and filtration systems, warning and backup systems could be eliminated."[14] Their paper also suggested that many safety requirements be liberalized in the areas of pressure suppression, pressure venting, and low-pressure containments.

The most interesting response to these suggestions came from the *Savannah* itself. One week after West and Roland advanced their money-saving proposal, the ship had to be brought back to its Hoboken pier after its instruments indicated a possible leak in the ship's secondary cooling system.[15] Whether there really was a leak, or whether one of the gauges was out of kilter, is a matter of conjecture. But the incident did seem to be telling us something about secondary safeguard systems.

There is no doubt that our merchant marine needs a shot in the arm. Whether it is desirable that this "shot" be administered by nuclear reactors is another matter. A nuclear fleet *can* be made competitive and economical *if* the taxpayer supports it with heavy insurance coverage and the high cost of safeguards; it can also be made competitive without taxpayer help if some of the safeguards are dropped and insurance premiums skimped. It cannot, however, be made economical *and* safe without something giving somewhere.

Even if the shipping lobby fails to get a nuclear merchant fleet launched, there will nevertheless be plenty of radioactive material on the high seas. It can take a number of forms: Partially refined uranium will be shipped to the United States by foreign countries for enrichment in gaseous diffusion plants; the enriched fuel will go back to the countries of origin for use in reactors; irradiated fuel will be shipped here for reprocessing; reprocessed fuel will be shipped back for reuse in reactors; and of course importation and exportation of radioisotopes for a wide variety of purposes will be a thriving form of trade.

Actually, shipping of nuclear material is already a respectable-size business. As of mid-1967, the AEC had distributed abroad through sale, lease, and deferred payment special nuclear and other materials valued at approximately $266.4 million, bringing in revenues of $171 million. Much of this was uranium 235—more than 10,000 kilograms of it, or about 10 metric tons.[16] In February 1966 President Johnson increased the allocation of enriched uranium for distribution abroad to 250,000 kilograms, or about 270 tons.[17] In 1967 legislation was enacted to permit the AEC to transfer to the European Atomic Energy Community 1,000 kilograms of plutonium in addition to the 500 kilograms previously authorized, and 145,000 kilograms of uranium 235 in addition to the 70,000 kilograms previously authorized.[18] This stuff will have to cross water to get to Europe, obviously.

The United States was also vigorously pursuing business in the area of enrichment of foreign fuel and reprocessing. So there will be lots of international comings and goings of radioactive material in the next few decades, with the odds that one ship out of three bearing it will be disabled—and one out of two hundred won't survive at all. Forty-four major American ports have been cleared for receipt of irradiated fuels.[19]

Of course, not all nuclear material goes abroad by sea. Some of it travels in planes, under NRC, International Commerce Commission, and Federal Aviation Agency rules. Domestic nuclear flights had been routine, but they have recently been curbed by the NRC. Until January 1976 most international nuclear cargoes passed through New York City's Kennedy International Airport. When New York banned these shipments, they were diverted to O'Hare, in Chicago, the busiest airport in the world. Mayor Michael Bilandic finally halted these shipments not quite two years later.[20]

New York City has been a pioneer in its healthy regard for the dangers of nuclear substances. Under the direction of Dr. Leonard Solon of the city's Department of Health, the city banned what had been routine shipments of radioactive material through the city's streets and highways, resulting in a precedent-setting court fight with the federal government in August 1975.[21] Waste from Brookhaven

National Laboratories on Long Island was then shipped by passenger ferry to New London, Connecticut, and then via truck to its southern resting place—until New London followed the example of New York a few years later. Many other localities have since instituted or considered restrictions of radioactive material transports. Meanwhile, accidents do happen. For example, a truck carrying radioactive cobalt lost part of its load along Route 80 in Pennsylvania in the fall of 1978.

Upsetting as these pictures of radioactivity in the air and on the sea may be, they form only two panels in our triptych of recklessness, callousness, and ineptitude. The third and center panel, featuring shipments by land, is a compelling study of human madness meriting comparison with the most tormented productions of Hieronymus Bosch.

Before gazing upon it, some background is necessary.

One of the fundamental rationales of atomic power is the cheapness of fuel transportation. Proponents of atomic power point out that a tremendous percentage of the cost of conventionally generated electricity paid by the consumer owes to the expense of bringing coal, oil, and gas by train, truck, boat, or pipeline to the utility. But because one ton of uranium delivers as much energy as about seventeen thousand tons of coal, uranium shipping charges are an inconsequential fraction of your electric bill. Time and again one reads such statements as the following, made by an AEC official in February 1968 before an American Public Power Association gathering: "The relatively minor cost of transportation of nuclear fuel provides greater flexibility in the location of generating plants and thus influences decisions with respect to long-distance transmission."[22]

Apart from the current trend for conventionally fueled utilities to construct power plants directly on mine and refinery sites wherever possible, thus sharply reducing fuel transportation costs, the AEC's transportation argument is grossly misleading in all that it fails to take into account.

First of all, uranium *does* travel all over the place. In *Atomic Fuel,* an AEC information pamphlet, a section endearingly entitled "The Odyssey of Uranium," informs us that "once the uranium has been separated from the ore dross, it is ready to travel, and travel it does. For example, material mined and milled in Utah may be refined in Missouri, enriched in Kentucky, converted in Pennsylvania, fabricated in California, used to generate power in Massachusetts, and reprocessed in New York!"

Second, it is never stressed that as the ore is refined, enriched, converted, fabricated, shipped to the user, and shipped back to the reprocessor, the radioactive content gets higher and higher, the attendant hazards steeper and steeper.

Third, no mention is made of the enormous expense involved in shipping the huge quantities of radioactive waste to storage sites, some of which are located in the State of Washington, three thousand miles west of Eastern Seaboard utilities. Special containers must be fabricated to contain this supremely toxic liquid against the contingencies of transportation, and other special precautions taken to ensure uneventful delivery to the turbulent caldrons in which it is to be stored for centuries. Ritchie Calder, in *Living with the Atom,* states that transportation of high-level and extremely dangerous radioactive wastes required trips of more than two million miles annually—and that was in 1962.[23]

Finally, when atomic power proponents talk about cheap transportation, little mention is ever made that much of the savings generated is due to the absence or inadequacy of insurance coverage. A study made by the Southern Interstate Nuclear Board found three major obstacles currently affecting transportation of radioactive materials. As reported in an AEC Division of Industrial Participation book, these obstacles are: (1) the refusal of the eastern railroads to transport fissile materials as common carriers; (2) the lack of satisfactory nuclear indemnification for movements on the high seas and by means of air transport; and (3) the lack of indemnification which, in the view of bridge, tunnel and turnpike authorities, would enable them to offer the use of their facilities as applied to other types of cargo.[24]

Thus the way to keep transport costs down is by not buying insurance. In the same AEC publication, one reads: "Air transportation of radioactive materials has frequently been the only solution to shipping cold nuclear fuels [meaning uranium sealed in metal tubes] and radioactive materials overseas because of the difficulty in obtaining indemnification for steamship service."[25] Another passage informs us that it cost about $2.50 and $3.00 per kilogram respectively to transport irradiated fuel from the Yankee and Dresden power plants to the Nuclear Fuel Services reprocessing facility at West Valley, New York. These prices included cask rental, but "insurance was not required in these instances." The NFS facility was where used fuel is divested of its intensely radioactive fission products. Thus violently lethal material traveled on public thoroughfares to New York from Illinois and Massachusetts totally without indemnification.

For anyone with a sense of black humor, the AEC's annual supplement to its Summary of Industrial Accidents have more entertainment value than the annual special edition of *Variety*. A typical harvest of transportation accidents involving release of radioactive material, reaped from the 1963–64 supplement, includes:

> During transit, plutonium-contaminated liquid leaked from a glass carboy (with a loose-fitting ceramic stopper) and contaminated several trailers, truck terminals, a number of packages of other materials and several persons handling the materials. . . . The 13-gallon glass carboy was in a large wooden box. The box was marked "this end up," so that during transit the carboy would remain upright. However, the presence of pallets on the box apparently suggested to the carrier that the box should be placed on its side, resting on the pallets, which allowed the material to leak slowly from the container and the wooden box. (The carboy was not surrounded on all sides by an absorbent material sufficient to absorb the entire liquid contents.) . . . Decontamination was carried out successfully in the truck terminals, trailers, and on the cargo at a cost of approximately $27,500. [Jersey City, New Jersey]

> A 55-gallon drum, containing 5 polyethylene bottles of radioactive material, was damaged during transit. . . . During transit, the ring bolts became loosened on two drums (lying on their sides) and the lids came off, allowing 4 bottles containing 93% enriched uranium to escape from one drum. [Kansas City, Missouri]

> During railroad yard switching operation, a railroad car door broke open and two drums, containing 0.947 enriched uranium, fell out. Part of the contents of one of the drums spilled. The entire shipment constituted 251 drums. [Madison, Illinois]

> Upon arrival at destination, an autoclave housed in a lined shipping cask and containing nine unclad fuel pins in 432 cc's of water was discovered to have leaked during transit. The leak occurred when a broken valve allowed contaminated water inside the autoclave to leak into the shield container and then seep out onto the floor of the truck. [Chicago, Illinois][26]

On top of these, the supplement described a large number of accidents in which no radioactivity releases occurred:

> Truck transporting fuel elements involved in accident with private car.
> Tractor-trailer carrying interplant shipment of radioactive materials tipped over when forced off road by oncoming traffic.

Train derailed. Coach carrying radioactive materials remained upright.
Truck transporting radioactive material jackknifed.
Truck carrying radioactive materials slid into ditch.[27]

That no releases occurred in these accidents is certainly no assurance that they will not occur in the future. All it means is that so far accidents involving radioactive material in transit have not been severe enough. But the laws of averages are working against the shippers. In April 1968, for example, it was announced by the National Transportation Safety Board that the number of train accidents had increased from 4,149 in 1961 to 7,089 in 1967—a zooming rise of 71 percent. Of the total number of accidents in 1966, 4,447 were derailments and 1,552 were collisions.[28] In a letter to the Federal Railroad Administration, the Board's chairman, Joseph J. O'Connell, Jr., stated that in the future the accident rate might get worse because of higher speeds, longer and heavier trains, and the *increased hauling of hazardous materials*. Nor have we considered sabotage. The aforementioned shipment of 251 drums of highly enriched uranium, two of which rolled out of their railroad car, would have made a compelling target for the saboteur's bomb.

More recently, accidents have been widespread. Kentucky, Pennsylvania, and Colorado have been among the states where mishaps have occurred. The worst so far dumped 15,000 pounds of uranium oxide, carried in ordinary industrial containers, onto a Colorado highway in 1977. Although it happened in a rural area, the truck's route ran through downtown Denver.[29] Also, many of these shipments use no more than ordinary security, making them prime targets—marked *Radioactive* as they are—for hijacking and other violence.

Considering the enormous annual increase in nuclear industry activities, there is a reasonably good chance that one day in the 1980s a truck or railroad car containing radioactive material will be blown sky-high or will dump its cargo in the midst of a town or city population, or in a drinking or sewage water system.

The inadvertent or intentional cracking open of casks containing nuclear material is just one eventuality we can expect if we permit the peacetime atom to come of age in the 1980s. Another is the loss or theft of such material. Yes, it *can* be lost. And yes, it *has* been stolen! Leo Goodman, who has fought a courageous battle to keep tabs on, and publicize, the misadventures of the atomic industry, stated in a

1967 speech that his tabulations showed eighty-eight capsules of radioactive fuel lost or stolen up to then.[30]

An instance of loss in 1966 received wide coverage in the newspapers. A one-inch-long cylinder containing three hundred milligrams of radium was lost in a shipment between Fort Worth, Texas, and Queens, New York. That the capsule had been poorly packed, enabling it to escape both from its lead container and the box in which the container was contained, was bad enough. But the real headache owed to the fact that authorities hunting for the capsule found it impossible to trace the route along which it had been shipped. "We are really disturbed," said a spokesman for the United States Public Health Service in Washington, "that the route of this hazardous shipment should be so hard to identify." Nevertheless, after monitoring a number of potential routes and terminals, searchers had to give up. The capsule is still — somewhere. In 1,620 years it will have lost half its radioactivity.[31] Six weeks later a second capsule was misplaced in much the same fashion.[32]

Sometimes lost radioactive material is found. In March 1962 a little boy named Henry Espindola, playing in a town dump outside of Mexico City, found a container and managed to pry it open. It contained seven pellets of radioactive cobalt apparently lost by a nuclear engineer some time earlier. The container was kept around the Espindola house as decoration. On April 17 the grandmother, Augustina Ibarra, noticed the drinking glasses in the kitchen darkening. By April 29 Henry had died. The mother died on July 19. The cause of death was recognized and the capsule removed on July 22. Nevertheless, Henry's sister died on August 18, and his grandmother on October 15.[33]

With increased movement of radioactive material along world traffic lanes, a new and incomparably frightening hazard poses itself: piracy. But what, you ask with a nervous laugh, can they do with a shipment of uranium — make a bomb with it?

An article on the front page of the June 13, 1968, edition of the *Wall Street Journal* may provide an answer. The article described a report prepared by a nine-member panel formed by the AEC in which it was predicted that a nuclear black market is likely to develop. "An atomic bomb can't be built without the right materials, of course," the article explained, "but atomic scientists say that as these materials gain increasing use, the opportunities for stealing them will multiply. Indeed, even now large shipments of nuclear material are regularly trucked in ordinary unarmed rigs through rural America, where hijacking would be relatively simple, and by 1980 these shipments will be sharply increased. Similarly, tremendous transatlantic air shipments of nuclear material for Europe's burgeoning nuclear power industry already are frequent."

The panel surmised that these shipments could attract criminal organizations that might divert the enriched uranium or plutonium. The once secret information needed to build nuclear bombs was unclassified years ago, and the *Journal* quoted Theodore Taylor, a nuclear physicist who headed the Defense Department's atomic bomb design and testing program for seven years, as saying, "I've been worried about how easy it is to build bombs ever since I built my first one."

Whether the matter is as simple as Taylor suggests is open to question, for the problems of handling, fuel separation, and so forth are probably beyond the financial or technical means of any criminal syndicate and undoubtedly beyond those of an individual criminal, revolutionary, or lunatic. They are not, however, beyond those of a nation, and the possibility that some countries could build—if they have not already built—secret installations for the purpose of producing atomic weapons from "diverted" material has been raised by reliable observers.

There is no question that workable rudimentary bombs can be *designed* by anyone with a common college-level knowledge of physics. Some years ago a Princeton student proved this point by submitting plans for an atomic bomb to the federal government. Moreover, "erroneously declassified" material was recently found *on the shelves of the public library* in Los Alamos, New Mexico. According to the *Washington Star,* these documents were available to all comers for four years.

And in Madison, Wisconsin, *The Progressive* magazine became the first victim of successful court-ordered prior restraint (prevented from publishing an article because of a claimed violation of national security) in U.S. history. In early 1979 the government discovered that the magazine planned to print an article on atomic secrecy by Howard Moreland and quickly moved to suppress it—this despite the government's admission that all the material therein was already available to any diligent researcher. Publisher Erwin Knoll expressed intense frustration because he was not even able to read the judge's decision; it too was classified as soon as it was handed down. After essentially the same material appeared in *The Press Connection* (another Madison, Wisconsin, publication) the government abandoned its prior restraint; it was lifted on September 17, 1979. *The Progressive* article finally appeared, half a year late, in the November issue.

The *Wall Street Journal* mentioned two incidents that had made a vivid impression on the members of the AEC-appointed panel. In one, two workers at a London plant stole twenty fuel element rods containing enriched uranium, dropping them over a fence for a pickup. The plot was discovered before the fuel elements were removed from the site.

In the second incident, a plant that makes fuel elements, reviewing its inventory, discovered that more than 100 kilograms of enriched uranium, totaling about 6 percent of the amount handled by the company over a six-year period, was missing. "At first," the account explains, "it appeared conceivable that the missing material had been diverted into bomb production somewhere. After a long hunt at the plant, however, a fraction of the material was found. The AEC then assumed that the rest was lost in normal processing—blown out vents, tracked out on shoes or buried as scrap." Delmar L. Crowson, director of the AEC Office of Safeguards and Material Management, stated that "we have no reason to believe the material got away from the plant," but some Federal officials, the *Journal* stated, were not quite as sanguine as Crowson.

There is no fully effective means of policing the peaceful atomic activities of our own or foreign nations. Indeed, the New York *World Journal Tribune* of April 16, 1967, pointed out that "nuclear reactors are sprouting so fast all over the world that it is almost impossible for the International Atomic Energy Agency to keep an accurate census of what is being built and what is planned." Thus the projected proliferation of civilian nuclear power plants presents mankind with the most bitter irony imaginable. For it may well turn out that the atomic weapon that triggers the next and final world war will be fashioned from fissionable material "diverted" from a power plant dedicated to the innocent purpose of producing electricity.

Advocates of Atoms for Peace have much to think about.

9

We Interrupt This Broadcast . . .

"We interrupt this broadcast . . ."—words guaranteed to accelerate the pulse and stay the breath. Fingers go to lips and conversation dies rapidly. Everyone leans forward and listens apprehensively, conditioned to expect bad news.

At first, the announcer's words leave his audience more puzzled than anything else: "An explosion occurred one hour ago at the nuclear power plant north of this city. Radiological experts are now considering the extent of damage, but city officials have been informed that evacuation of the surrounding area may have to be undertaken due to the possibility of radioactive contamination."

His next statement, however, would undoubtedly send hearts racing with terror: "A nuclear physicist interviewed by this station reports that under certain weather conditions, an area as large as a hundred miles in diameter could be affected, raising the possibility that the entire population of this city, as well as those of all communities within fifty miles of the reactor site . . ."

Then, inevitably, the words: "Residents of this area are advised to remain calm and stay tuned to this station for further bulletins."

Are you prepared for this announcement? Is your community?

Your city hospitals and health services? Your civil defense? Your state or federal government? From all indications, the answer is no: This situation would catch every sector of the public unprepared.

Clearly, no plan can be devised to cope, or even begin to cope, with a disaster of this magnitude.

The closest we have come was the Three Mile Island accident, and the official confusion and conflicting reports that followed are not apt to be quickly forgotten by the public, and, we hope, not by our more responsible representatives. It is particularly disturbing to realize, however, that the *need* for evacuation plans was recognized from the very beginning of the nuclear reactor program. A report dated December 1, 1953 stated, for example:

> In the past it has been the practice of the AEC advisory committee dealing with reactor safeguards to recommend the observance of large exclusion areas for reactors, the area in general determined by the quantity of radio-active and toxic poisons present in the reactor. Contrary to general belief, these extensive isolation areas do not preclude that the adjoining areas are free from the hazards of a reactor incident. Therefore, it was necessary for these committees to further recommend that the adjacent zones be prepared for evacuation in the event of a reactor catastrophe.[1]

More than twenty-five years have passed since that "recommendation." What is the current status of our nation's preparation?

According to one Associated Press report (May 11, 1979), sixteen states with operating reactors have no federally approved emergency plans: Colorado, Georgia, Illinois, Maine, Maryland, Massachusetts, Michigan, Minnesota, Nebraska, North Carolina, Ohio, Oregon, Pennsylvania, Vermont, Virginia, and Wisconsin.[2]

Another Associated Press report, also May 1979, noted that "Joseph M. Hendrie, chairman of the NRC, told a House Armed Services subcommittee that local governments are poorly prepared to deal with a nuclear accident that might require an evacuation."[3]

A report issued in 1975 by *Public Citizen,* "Evacuation Plans — the Achilles' Heel of the Nuclear Industry" by Ron Lanoue, raised several pointed questions with regard to the effectiveness of any evacuation plans that have been (or might still be) evolved. Among these questions:

> Will the roads carry the traffic of an evacuation? . . . For example, the Oyster Creek plant in New Jersey is located near an offshore resort island,

filled with tens of thousands of beach-goers in the summer. There is only one bridge linking the island to the mainland, creating the potential for a large, lethal traffic jam. . . .

Are there provisions for evacuating those without automobiles? . . . Check especially for schools, nursing homes and hospitals in potentially affected areas. . . .

Do these provisions (for public warning) appear adequate? For example, radio and television won't reach too many people at 3 A.M. . . .

How serious and how far-reaching an accident does the plan prepare for?[4]

The last question has particular significance. People reading the *Atlantic City Press* of January 11, 1977, saw the headline "N-Plant Guide Omits the Worst," and learned that the New Jersey state plan did not include guidelines for a worst case accident (a meltdown and breach of core and shell of reactor). Paul Arbesman, director of the State Division of Environmental Quality was quoted as saying, "We haven't planned for that simply because our government says it can't happen."[5]

A year later a test of the plan to protect Salem County (New Jersey) residents in the event of a reactor emergency produced results an *Atlantic City Press* editorial called "startling." The editorial continued:

Many local officials traced the basic problem to a lack of communications. It took one hour and 40 minutes for township officials to be notified that a simulated disaster was being conducted at the generating plant. The plant is less than five miles from the new township hall.

Officials in neighboring communities, including some municipalities that were never notified at all, expressed concern over the test. . . .

"We'd be in terrible shape," according to Mayor Carelton Sowers. "There's no organization, they never set up anything."

Public Service Electric and Gas officials weren't as upset, and some state officials say they were satisfied with the results of the test.

The implications are something every citizen in every part of our country should ponder.

Whatever plans the State of New Jersey has made, however, have not included adequate hospital preparation to deal with less serious accidents. In a letter to the authors from Dr. Willard W. Rosenberg, M.D., of ACCCE (Atlantic County Citizens Council on Environment), dated May 19, 1978, about an encounter at a public debate on the issue of nuclear safety:

> My question, put to Smith [president of PSE&G] and Bartnoff [president of Jersey Central Power & Light Co.] was: Since there is no hospital or other medical facility in New Jersey or any known adequate medical technology to deal with even a moderate radioactive catastrophe, much less a major one, how can you morally continue to promote nuclear power plants in New Jersey? Smith said he thought the money PSE&G gave to a hospital in Salem was for that purpose. I assured him the Salem hospital could not do the job. Barnoff said it was an emotional question and, despite my request that he answer the question, he spoke of brown-outs and people freezing in winter if there were no nuclear plants. At this, Lovins [Amory Lovins of Friends of the Earth] remarked, "Who's being emotional now?"

In view of the accident at Three Mile Island, the following testimony given July 26, 1974, to the ACRS's Environmental Subcommittee has ironic significance:

> *Harold Collins* (Office of Govt. Liaison, AEC): Now, in answer to your question, are there any states that have 100 percent totally acceptable radiological emergency response plans, the answer to that is no.
>
> *Dr. Isbin*: Not even Pennsylvania?
>
> *Mr. Collins: Not even Pennsylvania. Because Pennsylvania, you will remember, is a model state plan that supposedly is under development.* [Emphasis added.]

Reprinted in the back of Lanoue's article is a letter dated April 8, 1975, from William Kerr, chairman of ACRS to William A. Anders, chairman, U.S. Nuclear Regulatory Commission. The subject was emergency planning, and it included the following:

> On the basis of its evaluations, the Committee has concluded that an effective emergency plan can play a significant role in the protection of the nearby population in the unlikely event of a major accidental release of radioactive material from a nuclear installation. Reviews by the Committee, however, of emergency plans for nuclear power plants currently in operation, or nearing completion of construction, show that *much work remains to be done.*[9]

"Much work remains to be done." The years pass. The rhetoric remains almost the same, word for word. Nuclear plants are built, operate for a time, break down or are shut down for long periods to correct newly discovered design deficiencies or other problems — and at great expense to the consumer, to whom the repair bills and cost of replacement power are inevitably charged. The public relations efforts of the nuclear industry and other proponents increase and yield their full harvest of public confusion.

And a good deal of the most important work of all *"remains to be done."* For example, the work of preparing hospitals and their personnel to deal at least with transportation accidents involving radioactive materials; the work of preparing firemen and police to cope with such accidents (and those at reactor sites); and the work of providing the public with an evacuation plan and information on best possible procedures in emergencies, which are, as we have seen, grossly inadequate *at best.*

Nothing, in fact, has changed (except that the hazards have grown larger and more numerous) since the 1963 symposium on radiation accidents and emergencies, at which one panelist stated: "Take a transportation accident that happens in the middle of a cornfield somewhere. The nearest hospital has got to have some sort of preparation for taking care of this . . . I know several radium installations not too far from here where I'm sure there has not been the slightest preplanning for what would happen if there was a spread of contamination."

It is doubtful, that even with much broader training programs, we will ever be able to prepare for radiation accidents involving several hundred victims, much less for one involving injury to thousands, tens of thousands, or hundreds of thousands. Nevertheless, it is imperative that the public and its legislative representatives recognize that the current programs of the NRC and the U.S. Public Health Service hold little promise against the possible harm in an accident of *any* serious scale.

One of the biggest problems that would arise in the case of such an accident is that of decontamination. Unless medical personnel were completely informed and trained, plentiful equipment at the ready, and a well-thought-out plan drawn up and impressed on the minds of regular hospital staff and volunteers, the initial catastrophe could be multiplied by hysteria and gross confusion as nonirradiated individuals did their best to avoid being contaminated by victims rushing to them for aid.

It should be noted that people may be injured by radiation without actually contacting radioactive material or debris. Radiation sickness is not contagious; one person cannot "catch it" from another. On the other hand, when the victim has been directly exposed to fallout —very small particles of radioactive material attached to dust and other matter suspended in the air—the victim himself, along with his clothing, is a "carrier," capable of irradiating other persons. Clothing must be removed and decontaminated and the victim washed down to remove radioactive dust from all skin surfaces; otherwise he can expose countless other people.

The complications are many and difficult to overcome. For instance, referring to the room to which victims are originally taken, Dr. Robert Landauer observed:

> If the room is connected with a general air conditioning system means must be provided for a "disconnect" to prevent the dissemination of radioactive material throughout the hospital. It is important always to remember to "contain" radioactive material, not to spread it. This applies to the decontamination team personnel as well. These people should not leave the area without first ascertaining that they themselves are not contaminated. There have been several instances of physicians spreading radioactivity all over a hospital and even into their homes.[6]

Another fearful complication arises when one considers the procedure of washing down the bodies of contaminated persons. For while it is feasible for minor transportation or nuclear plant accidents in which radioactivity does not escape into the environment at large, it would be far less feasible, and perhaps impossible, in the vicinity of a large nuclear plant accident where fission products and gases were released outside the utility structure—*because the water supply itself might be contaminated.* Bear in mind that nuclear plants must be located on the banks and shores of large water sources for cooling purposes, and those water sources invariably supply drinking and bathing water to nearby communities.

As H. M. Parker and J. W. Healy pointed out several years ago:

> Contamination of bodies of water will occur both by direct fallout and by secondary leaching of the materials into the streams. Direct fallout could render bodies of water reasonably close to the reactor unfit for use until the material is carried away. . . .
> The possibility of rainout into a stream near the reactor site exists, with

severe contamination resulting, depending upon the rate of washout from the cloud and the stream flow characteristics. Of more concern in at least some cases is the possibility of the escape of reactor coolant containing a significant quantity of fission products. Such a mishap would bring about a band of grossly contaminated water, depending again upon the flow characteristics of the stream, which could cause severe contamination of water plants or other equipment downstream and in special cases could lead to serious radiation dosages to people using the water for sanitary purposes.[7]

More recently, writing about the effects of a nuclear attack, a disaster that would not differ substantially in its effects on water supplies from a major nuclear plant release of radiation, Tom Stonier stated in *Nuclear Disaster*:

Water supplies may also become temporarily unusable as a result of heavy fallout contamination. In Westchester County [a suburb of New York City], for example, about 90 percent of the population uses water from open surface sources, and if fallout were fairly heavy, this water would not be fit even for emergency consumption for days or possibly weeks. It would probably be acceptable for emergency use thereafter, because fission products in fallout appear to be in an insoluble form, and tend to be removed by sedimentation and filtration. However, even this "uncontaminated" water might not be entirely safe, since it could still produce considerable long-term biological damage, particularly from radioactive iodine.[8]

It must also be remembered that in the event of a major reactor accident, hospitals and clinics in the area might themselves be severely contaminated and require evacuation. In fact, a moment's thought produces the realization that scores of fundamental questions remain unassessed. For instance, what if the accident occurs late at night, catching most of the public asleep? Some communities might be able to resort to air-raid sirens and the like to alert citizens, but is there any assurance that that measure would be taken, or that its meaning would be comprehended? How many cities have established evacuation routes for civil or military emergencies, to prevent the rampant hysteria that would quickly develop when mobs poured into main arteries? Without organization and discipline, a panicky citizenry could inflict dreadful injury on itself even if a contamination hazard were not present.

What would be the reactions of surrounding communities to the prospect of countless numbers of fleeing, and possibly contaminated, individuals? It has been pointed out, for example, that during the plague that struck London in 1665, surrounding towns padlocked water wells, closed shops, and posted armed guards on the roads from the city to bar fleeing refugees. And in the 1950s, when fear of nuclear war ran highest in this country, it was common to hear talk of means of repelling survivors from affected areas, slaying them in self-defense or in protection of food, medical supplies, and so forth.

Another possibility unconsidered is that people living near other nuclear reactors—and of course there will scarcely be a population concentration in this country *not* within reach of radiation releases— may very well become infected with panic, creating disorders throughout the land. Operators of other power plants might become jumpy and detonate yet other accidents, a possibility that many psychologists, examining the mysterious clusters of accidents occurring from time to time, such as airplane crashes, believe to have a substantial basis in human behavior. Our government might have to call for a temporary shutdown of all atomic plants—especially if there is suspicion of sabotage—pending investigation or until the first wave of panic subsides. In an age when one-fourth or one-half of our electricity will be generated by nuclear energy, as has been forecast, what effects would these massive shutdowns have on the economy and the public welfare?

The fact that this country's fire, police, medical, and governmental personnel are inadequately prepared for a reactor plant or nuclear transportation accident of any magnitude, and its citizenry is utterly in the dark about even the most elementary do's and don't's in protection of self and property, in recognition or detection of radioactive contamination, in medical or evacuation procedure, is one of the most reprehensible aspects of the atomic power program. And it explains why it is so difficult for the average person to get any information about what to do if there is a serious incident. To illustrate this one need only to get some information from the local Civil Defense Office on procedure in such an event.

An acquaintance of one of the authors, for instance, called New York City's Civil Defense Office and asked what she should do if a serious nuclear plant accident occurred. She was told to "stay where you are." Assuming this instruction meant for her to hold on while the clerk checked with a superior for further directions, the caller waited. Some time later, another clerk picked up the phone and asked: "Are you waiting for something?" The lady explained she'd been told to stay where she was. The clerk's response nearly floored her: "Well, those *are* the instructions: Stay where you are."

Even conceding that this meant to remain in place until further instructions could be broadcast to the public, it hardly answers the question. "Where she was" could have been a phone booth.

A caller today would probably be told that "the NRC would step in." A good many people living in Pennsylvania and surrounding states did not find that particularly reassuring in March and April 1979.

10

The Thousand-Year Curse

"How fortunate it is that our discovery of radiation is paralleled by our knowledge of its deleterious effects. . . . Otherwise, ignorance of the consequences of irradiating man might induce great reckless-ness."[1] The passage is taken from *Radiation* by professors Jack Schubert and Ralph E. Lapp.

There is scarcely anything in our histories to reassure us that man is capable of conducting himself wisely in the presence of known lethal hazards. Indeed, Schubert and Lapp portray instance after instance of human carelessness with radioactive emitters despite the abundance of well-documented dangers. One can only assume that when they wrote the poignantly wishful passage, they were desperately hoping that in the future man would have learned enough to proceed with greater caution in coping with the unprecedented hazards of radiation.

Thus far this book has dealt with the possibilities and potential consequences of disasters at reactor and other atomic facilities or in transit. It must be recognized, however, that even if such massive emissions could somehow be prevented, our population will be exposed to a sharply increasing amount of radiation that the government and atomic power industry foresee as the unavoidable byproduct of tomorrow's peacetime atomic activities. Disturbingly, these emissions are labeled "planned releases." Furthermore, stored in fragile and unsafe tanks will be hundreds of millions of gallons of savagely toxic nuclear waste material, seething with many times the amount of radiation necessary to destroy all life on earth. The technology for neutralizing this radiation is unknown and, even to the best scientific minds, practically unimaginable; the technology for safely disposing of it is scarcely well enough advanced to make a difference in the crucial decades ahead; and the technology for merely containing it is so studded with unknowns and hazards as to make it utterly unreliable.

Despite all preventive efforts, future atomic facilities will be releasing enough radiation into our environment to present a clear-cut danger to every living thing on this planet. Ritchie Calder, in *Living with the Atom,* describes an "audit" of environmental radiation that he and his colleagues, meeting at a symposium in Chicago, drew up to assess then current and future amounts of radioactivity released into atmosphere and water.[2] The audit covered the period 1955–65 (the book was published in 1962), and because atomic power was negligible then compared to the prospects for the 1970s and beyond, the high figures on the ledgers are most significant.

The members of Calder's group assumed that in the period 1955–65, "the figures for land-based power reactors might be a thousand or ten thousand megawatts of heat which, in turn, means that locked in the total system there would be a thousand or ten thousand million curies of radiation. And the 'planned release' (the acceptable amount of 'leakage,' given all precautions) would be a millionth of that." Of course, "a millionth of that" comes to 10,000 curies per year. By 1980, Calder estimated, the annual leakage would be about 100,000 curies. This figure did *not* allow for radioactive releases resulting from serious reactor accidents, however.

Nuclear ships were then considered, including both military vessels—more than thirty nuclear submarines were under construction at that writing—and commercial, such as the N.S. *Savannah.* The symposium team estimated that 60 million curies of radioactivity would be present on or under the high seas in ship reactors, and "planned release" would come to 3,000 curies.

Nuclear aircraft propulsion, which at the time of the symposium was still being seriously weighed by our government, was regarded by the "auditors" as capable of releasing at least one million curies annually into the atmosphere, and by 1980 the figure could rise as high as a hundred million. Fortunately, that area of research and development was dropped, though nuclear propulsion of rocket ships is a present field of investigation. Also, reactors were being designed and used to provide power for instruments and radio communications in satellites. "If we have a lot of megawatt auxiliary power units in space vehicles by 1980," Calder wrote, "flying all around outer space but occasionally entering the atmosphere as live reactors, we might have something of the order of a million curies of fission products introduced into the environment."

The audit examined research and test reactors (as opposed to electricity-generating commercial reactors). "In the United States, these already represent something like 1,000 megawatts and, therefore, 1,000 million curies. Of these about a hundred thousand curies escape every year," Calder stated. He continued:

> There are the plutonium factories. These release into the environment something like a million curies a year, in the form of low-level wastes. . . . The uranium mills seem to be releasing about 1,000 curies of radium per year into the rivers of the United States. . . . There are also the fuel reprocessing plants. There are the radioisotopes which are being distributed all over the world for research and industrial uses and for medical treatment. These are under control, but by 1965 they will represent 10 million curies. The absolute safety of these depends upon the responsibility and training of those handling them. They are stored in impenetrable vessels but sometimes they get loose.

The final tally made by Calder's group was most disquieting: "By the time we had added up all the curies which might predictably be released, by all those peaceful uses, into the environment, it came to about 13 million curies per annum."

Calder went on to say, as if to belittle this figure, that several thousand times that number of curies had been released into the atmosphere in each of the five years of active testing of nuclear weapons prior to 1962. But that assertion not only does little to comfort us, it actually raises even sharper anxieties, for it means that the unnatural radiation in our environment was already extremely high by the time peacetime uses of nuclear power began releasing radioactivity.

The 13-million-curie annual "planned release" figure included

nuclear aircraft propulsion, which currently is not a serious consideration and can be discounted. On the other hand, the audit did *not* include serious reactor accidents; it did *not* include experimental programs in which some radiation inevitably would be released, such as
the "Plowshare" program of AEC-conducted explosions exploring the
use of atomic energy to excavate harbors and canals; it did *not* include potential accidents involving transportation, on land, on sea,
and in the air, of nuclear materials; it did *not* assess the contribution
of radioactive fallout through bomb testing, testing that still is carried
out today by nonsignatories of the Nuclear Test Ban Treaty. And it did
not include possible escape of stored high-level radioactive wastes,
the implications of which were awesome to contemplate: "What kept
nagging us," Calder said, "was the question of waste disposal and of
the remaining radioactivity which must not get loose. We were told
that the dangerous waste, which is kept in storage, amounted to 10,000
million curies. If you want to play 'the numbers game' as an irresponsible exercise, you could divide this by the population of the world
and find that it is over 3 curies for every individual."

Irresponsible or not, Calder's "numbers game" gives a clear indication of how much lethally radioactive material must be contained
without mishap; just as important, it indicates the impressive amount
of radioactivity that is being and will be released under so-called controlled conditions even if no mishap occurs—a most dubious proposition.

A detailed look at the ways in which the environment is contaminated by normal functioning of nuclear power plants and related facilities is enlightening. In a typical reactor, as we have seen, the concentration of uranium fuel in the core produces intense heat to convert
water into steam, which in turn powers electricity-producing turbines.

At the same time that the uranium is generating heat energy,
however, it is generating another form of energy, radioactivity, the
emission of rays such as beta, gamma, and X rays. The radioactivity
has a number of negative effects on the material inside the reactor.
One of these is swelling, warping, or destruction of the metal tubes
containing the uranium. Another is irradiation of the reactor components. Trace impurities in the metal cladding of the fuel tubes, for instance, or in the coolant, become radioactive, and once they do they
interfere with the efficient energy production of the uranium. After a
while, the accumulation of these "fission products" makes it impos-

sible to keep the reactor running economically. Also, the neutron bombardment of fertile uranium 238 by fissionable uranium 235 is creating plutonium, and it is desirable to separate that plutonium from time to time and utilize it as weapon or commercial fuel material.

Thus because some fuel tubes may be damaged, and because the uranium fuel has become contaminated by fission products, and because it is desirable to recover the plutonium byproduct, the fuel in the reactor's core must be replaced periodically. It is interesting to note that replacement occurs when only a tiny fraction of the fuel has been burned up.

The intensely radioactive spent fuel elements are removed from the reactor and stored under water for several months to allow some of the radioactivity to diminish. Then they are loaded into heavily shielded casks and shipped to a fuel reprocessing plant. There, by means of remote control equipment, the metal cladding is stripped from the fuel, the fuel dissolved in acid and the acid solution subjected to chemical separation processes. During these cycles, the noxious fission products are first removed from the fuel, then the uranium and plutonium are separated from each other and processed for reuse. The highly radioactive solutions from the extracting process are collected and boiled down to reduce their volume, and the concentrate is stored in large underground tanks.

Described in this abbreviated form, the whole process sounds sensible, safe, streamlined, and sterile, which is exactly what proponents of nuclear power would like us to believe. Pamphlets put out by government and industry describing the magical odyssey of uranium are replete with photos of white-clad workers in aseptic surroundings, cheerfully carrying out their tasks as if they were packaging candy bars. A closer look at these processes, however, is nowhere near as soothing.

Taking reactor operation first, it must be pointed out that the technology for trapping every last radioactive particle released by the uranium fuel is either unperfected or too costly. While most of these contaminating radionuclides are trapped by processing the air and water, then concentrated and shipped to storage areas, minute quantities are issued into the environment. These are known as "low level" wastes, and we are told that such releases are undertaken in such a way as to ensure dilution sufficient to prevent any predictable human exposure above levels expected to produce deleterious consequences.

Gaseous wastes are sent through a stack presumably high enough to ensure atmospheric dispersion. Liquid residues are spilled into the river, lake, or sea and presumably diluted to the point of harmlessness.

Thus when reactor operators are asked about the dangers of contaminating the environment, they cite the above techniques and the seemingly tiny amounts of radioactivity involved as evidence of the complete absence of risk in nuclear power technology.

These arguments are seriously misleading. As will presently be detailed, (a) a number of radioactive waste products released at reactor sites do not decay for dozens or even hundreds of years, which means that the accumulation of radioactivity in our environment from *all* sources will increase; and (b) many of these released radioisotopes tend to concentrate in human and other organic systems, eventually building up formidable doses of what Walter Schneir, in *The Reporter,* so aptly described as "the atom's poisonous garbage."

The second major aspect of the atomic waste problem comes in the reprocessing of the fuel used in reactors. Until 1966 the task of reprocessing spent fuel was strictly a government affair, having mostly to do with plutonium for defense purposes. The small volume of civilian reprocessing business did not justify the existence of a commercial reprocessing plant. But it is predicted that by 1980 there will be more than three thousand tons per year of spent fuel to be reprocessed, and by the year 2000 that figure will jump ten times, to thirty thousand tons.[3]

In anticipation of the vast increase in nuclear power in the 1970s, W. R. Grace & Co. and American Machine & Foundry Company formed an 80–20 partnership to erect the world's first commercial fuel reprocessing plant, sited about thirty miles south of Buffalo, New York. This $32-million Nuclear Fuel Services, Inc., plant has a capacity of 1,000 kilograms of fuel a day, about a quarter of a ton, and has an arrangement with the New York State Atomic Research and Development Authority for related radioactive waste storage facilities.

In December 1966 *Nucleonics,* a nuclear industry journal, chirped, "The world's first commercial fuel-reprocessing plant handled its initial 100 tons of spent fuel in a smooth and routine fashion — a remarkable accomplishment for a facility relying largely on remote operation."

This smoothness of operation, however, was not long-lived. The West Valley plant was one of the most troubled in the nuclear industry's history. In its six years of operation, it did process 625 metric tons of waste, but it did not operate at a profit. Many workers were contaminated, and the plant was eventually abandoned by its owner, Getty Oil. The company claimed the $600 million needed to bring the plant into compliance with new regulations was not worth it. The plant was closed in 1975.

Meanwhile, the state of New York has been left with the dubious honor of custodianship. New York is now stuck with 600,000 gallons of high-level waste so hot that it boils spontaneously — and all this on a site that is seismically active. It will cost $1 billion to decommission that plant; New York is already spending $2 to 3 million per year just to maintain the wastes in their present state. Furthermore, these dangerous substances have leached into nearby water sources, including Cattaraugus Creek and Lake Erie.[4]

During the plant's career, an actively opposing role was played by the Radioactive Pollution Subcommittee of the Rochester Committee for Scientific Information. The subcommittee arranged for one Wayne Harris to collect water samples from Buttermilk Creek below the NFS plant.[5] Buttermilk Creek is a tributary of Cattaraugus Creek, which forms a reservoir at Springville, then continues on to Lake Erie. The group had the samples analyzed, and on February 24 and 28, 1968, the RCSI published its findings. We mention here only those for strontium 90, the long-lived bone-seeking isotope, though other isotopes were reported as well.

The maximum limit of soluble strontium 90 allowed by New York state and federal law is three hundred picocuries per liter. On December 16, 1967, the committee and the New York Health and Safety Laboratory of the U.S. Atomic Energy Commission, measuring the amount of strontium 90 at the NFS sewage outlet, found a concentration of 86,900 picocuries per liter. This would amount to some 290 times the legal limit. The amount of strontium 90 measured in Buttermilk Creek on that date was 4,010 picocuries per liter, or about thirteen times the legal limit.

The average concentration of strontium 90 in the creek between December 1966 and May 2, 1967, was 655 picocuries per liter, or twice the legal limit. These findings would seem to confirm a New York Department of Health Radioactivity Bulletin issued in February 1967, in which "gross beta levels" of radioactivity in water downstream from the NFS plant, measured in November 1966, ran between 26 and

328 times the average measured at other water stations in the area.

The U.S. Congressional Joint Committee on Atomic Energy had criticized the AEC and Nuclear Fuel Services for inadequate attention to certain safeguards. Now Buttermilk Creek was adding its testimony. "The Nuclear Fuel Services Plant," the Rochester Committee concluded in its report, "is currently operating at a fraction of its licensed capacity. . . . If the plant has not managed to clean up its outflow after a year under minimum load, the RCSI sees reason for serious concern about its performance under full load."[6]

And well it might.

The reprocessing end, too, has proved to be highly unprofitable for the nuclear industry, despite the skyrocketing cost of uranium. General Electric spent $65 million to build a reprocessing plant in Morris, Illinois. When, in 1975, recurrent problems showed up during initial testing, GE ditched the plant; it has never been, and will never be, used.[7]

The only other commercial reprocessing plant in the United States is in Barnwell, South Carolina. Massive federal bailouts have kept this plant from being abandoned, but it has been a major target of antinuclear protestors, several hundred of whom have been arrested at the site. The site is now being used for waste storage, but not for reprocessing. Interestingly, the operators of Barnwell refused shipments of nuclear waste from the Three Mile Island plants after the accident there, claiming the waste was of a level far higher than they could deal with. The waste then had to be shipped to Washington State for storage.

It would be a pleasure to report that the worst part of the story had been told. But a survey of the perils of nuclear peace is like a Dantean survey of hell: One leaves each level certain that he has seen the worst, but ahead lie scenes more fiendish still. In order to complete our picture of the threat of radioactivity in the environment, we have to descend into that ring of Hades called Waste Storage.

It has been estimated that a ton of processed fuel will produce anywhere from forty to several hundred gallons of waste. This brew is a violently lethal mixture of short- and long-lived isotopes. It would take five cubic miles of water to dilute the waste from just one ton of fuel to the maximum permissible concentration. Or, if we permitted it to decay naturally until it reached that safe level, consider this: Just one of the isotopes, strontium 90, growing feebler by half every 27.4

years, would still be too hot to handle 1,000 years from now, when it will have only one seventeen billionth of its current potency. There is no way to reduce the toxicity of these radioisotopes; they must decay naturally, meaning *virtually perpetual containment.*

Unfortunately, mankind has practically no experience in perpetual creations, and the procedures for radioactive waste containment therefore leave a lot, if not everything, to be desired. Between 1945 and 1960, according to a speech made by Dr. Joseph A. Lieberman, chief of the AEC's Environmental and Sanitary Engineering Branch, some 14,000 curies of radioactive waste were dumped into the Pacific, and 8,000 into the Atlantic, contained in very much less than perpetual metal casks.[8] In 1959 Herbert Parker, manager of the Hanford Laboratories of the AEC, told the congressional Joint Committee on Atomic Energy that some three billion gallons of intermediate level wastes had been admitted to the ground at Hanford.[9]

The most common practice, however, is to store the concentrates in large steel tanks shielded by earth and concrete. This method has been employed for some twenty years, and more than seventy-five million gallons of waste is now in storage in some two hundred tanks. This "liquor" generates so much heat that it boils by itself for years. Of course, this satanic mixture must be kept constantly under surveillance; if allowed to boil dry, it would immediately begin eating its way through the tank and down to the water supply.

This task must be maintained essentially forever. Let's examine just one of the hundreds of nuclear byproducts that must be thus guarded: the aptly named plutonium. Large, modern reactors produce about 500 pounds per year.[10] With seventy-one operating reactors, a few of which are older, smaller models, we are probably producing more than 20,000 pounds *every year* (we manufactured 12,000 pounds in 1975). Estimates of potential cancer cases (in smokers) caused by *each pound* of plutonium run as high as 42 billion—a sobering thought considering that the current world population is fewer than four billion.

About 10,000 years ago human beings came to establish cities and farms along the fertile river valleys; King Tut's treasures—made of such inert materials as wood and stone—have survived miraculously for 3,000 years. After 24,000 years a pound of plutonium created in 1979 will have the toxicity of a half pound of plutonium—when one-thousanth of a gram (22 millionths of a pound) is clearly acknowledged as a surely fatal dose.[11] In 50,000 years there will still be a quarter-pound's worth of poison in that pound.

Each ton of plutonium will still have a lethal one-thousandth of a pound (2.2 grams, or at least 2,200 fatal doses) in 500,000 years—far longer than recorded history. It follows that our annual plutonium output today of ten tons for just one year of commercial nuclear power generation, with under a hundred operating plants (projections once went as high as 1,000 operating plants in the United States by the turn of the century), will have as much lethal radiation in a half-million years as a hundredth of a pound does now. It could still cause 420 million cancers in smokers, and a somewhat lower number in non-smokers.

Earthquake, flood, fire, change of government, terrorism, or just the ordinary ravages of time must not be allowed to affect stored wastes. Yet, as we shall see, the record so far is not one to inspire confidence in infallibility.

So far, we have been storing wastes generated since the World War II era, an infinitesimal fraction of the total storage time. Although thousands of pounds of radioactive wastes are generated each year, no permanent solution has been found. Meanwhile, many of the interim storage sites have had their problems, including the worst nuclear accident so far.

In 1958, according to dissident Soviet scientist Zhores A. Medvedev, who revealed this information to Western news media in 1976, a nuclear waste dump in Kyshtym, in the Ural Mountains, suddenly exploded, scattering radioactive clouds over a huge area, killing hundreds, and contaminating thousands. People showed up at local hospitals with skin sloughing off exposed areas. More than twenty years later, "the area . . . is now a wasteland. Drivers are warned not to stop as they pass through. Villages lie burned and abandoned."[12]

While the United States, which knew of the Soviet accident many years before the information was released to the American public, has so far avoided outright calamity, our waste storage sites have had their share of mishaps.

• Waste storage facilities at the huge federal nuclear complex at Hanford, Washington, have leaked at least eighteen times, releasing 430,000 gallons of high-level waste.[13]

• A trench at that same Hanford reservation was filled with so much plutonium that officials feared a heavy rain would separate and concentrate it into a critical mass—one that would sustain an atomic chain reaction. This could have caused an explosion similar to the Russian disaster. At one point, the AEC asked Congress for $1.9 million to mine plutonium out of the trench, known as Z-9, while at the same time filling a similar trench, Z-18, with 1.5 million gallons of plutonium waste![14]

• Water sources have been contaminated at Maxey Flats, Kentucky, Oak Ridge, Tennessee, and elsewhere, as a result of waste seepage.[15]

• Permanent disposal plans for a salt mine in Lyons, Kansas, were abandoned because the area was not sufficiently geologically stable; ground water would have been contaminated.

• Four storage tanks have failed at Aiken, South Carolina, along the Savannah River.[16]

Until recently, nuclear weapons accounted for nearly all of the nuclear waste now being stored. In recent years, however, the civilian nuclear power program has begun to generate a fearsome amount of these poisons. Each 1,000-megawatt reactor annually produces 9,000 liquid gallons and a hundred cubic feet of solids—and this is just the high-level waste![17]

It is not only the volume that fills one with apprehension, but the ugly disposition of this material. David Lilienthal put his finger on the crux of the matter when he stated:

> These huge quantities of radioactive wastes must somehow be removed from the reactors, must—without mishap—be put into containers that will never rupture; then these vast quantities of poisonous stuff must be moved either to a burial ground or to reprocessing and concentration plants, handled again, and disposed of, by burial or otherwise, with a risk of human error at every step.[18]

It cannot be stressed strongly enough that we are not discussing days, months, or even a few years. We are talking about periods "longer," in the words of AEC Commissioner Wilfred E. Johnson, "than the history of most governments that the world has seen." Nor can it be overemphasized that we are not talking about typical industrial pollution, requiring millions of gallons of waste to befoul a lake or river. We are talking about a material so potent that a few gallons released in a city's watershed could cause death and chaos on a scale comparable only to the havoc of modern warfare.

Bearing these things in mind, would we not be safe in assuming that safeguards no less fanatic than those surrounding the release of nuclear missiles would be operative in the case of nuclear waste management?

We would not.

In an article entitled "Radioactive Waste Management," published in the January 1965 issue of *Reactor Technology,* an AEC publication, W. G. Belter and D. W. Pearce cited no fewer than 9 cases of tank failure out of 183 tanks located in Washington, South Carolina, and Idaho. More disturbingly, a passage in the AEC's Authorizing Legislation for 1968 called for funding of $2,500,000 for the replacement of failed and failing tanks in Richland, Washington:

> A total of 149 tanks with about 95 million gallons capacity have been built [the report stated]. Prior to 1964, five tanks have been withdrawn from service because of leaks. Since 1964, one tank has failed and four more have developed indications of incipient failure. . . . The waste storage situation has been further aggravated by recent temperature control problems which have occurred in a tank being filled with the Purex self-boiling wastes. The tank bottom temperatures began exceeding the design control limits and further filling of this tank has been stopped which reduces planned waste storage space.[19]

The urgency of the situation could not be disguised by the bureaucratic language in which the plea was couched:

> There is no assurance that the need for new waste storage tanks can be forestalled.

Here then are tanks that have failed after twenty years. Twenty years is a respectable period as industrial processes go, but in the life of a radioisotope it may be but a moment, for some isotopes, as we have seen, can outlast ten, twenty, fifty, or more transfers of waste over hundreds and hundreds of years to come. It is beyond belief that this burden of centuries can be borne without some collapses. The individual tanks require elaborate supporting facilities. Special cooling apparatus is needed, itself subject to heavy contamination. Special pumps are needed to recirculate the waste to prevent ferocious surges of boiling. Switching and access equipment is needed to direct or divert flow from one tank to another.

Furthermore, these so-called tank farms are vulnerable to natural catastrophes such as earthquakes, and to man-made ones such as sabotage or plain human error. And what about the long-term legal and financial responsibilities? In 1959 the congressional Joint Committee on Atomic Energy, examining industrial radioactive waste disposal, predicted a capital outlay of $200 million for these atomic sepulchers, plus $6 million a year to maintain them.[20] Cannot private companies fail, or government administrations withdraw support for a program?

One needn't look into a crystal ball for a glimpse of what could happen if someone failed to hold up his end of the bargain in the matter of waste maintenance; it has already happened. On February 2, 1967, the *St. Louis Post-Dispatch* reported that a mining company had defaulted on an AEC loan to buy some 100,000 tons of uranium ore residue located on a field in Missouri's St. Louis County. It therefore became necessary to put the stuff up for auction—*repeat,* put it up for auction! It was picked up by a loan corporation and was still waiting for someone with sufficient financial backing to take it over and reprocess it. Title to 100,000 tons of radioactive waste—kicking around like a pawnshop ticket.

Efforts are of course being made for the effective handling of the waste problem. The most promising proposal would reduce the wastes to solid form, enclose them in concrete or glassy material, and store the blocks in abandoned salt mines proved to have no access to underground water systems. But while this is nice in theory, many technological barriers remain before the problem is licked. It is easy for AEC Commissioner Wilfred Johnson to toss off, "The Commission is now looking at the challenge of long-term waste management and I am certain we will have effective means in hand to meet it well in advance of the need."[21] But his optimism has no basis in solid fact.

Meanwhile, millions and millions of gallons boil furiously inside their frail tanks, tanks whose seams groan under strains metal was never meant to bear, and men go on building power plants to feed ever more of this corrosive ichor into the nuclear garbage dump.

Industrial waste is not peculiar to the nuclear industry. It is a phenomenon of every major industrial process. But the waste of the nuclear industry is unique in three ways. It cannot be disposed of in any conventional sense of the term. Its toxicity is not immediately apparent to human senses. And the longevity of certain waste isotopes poses storage and disposal problems unparalleled in industrial history.

These unique circumstances place intolerable burdens on government regulatory agencies and on the individuals in private industry whose task it is to supervise the waste management operations; and the load they place on future generations is unmitigatedly cruel. Joel A. Snow, writing in *Scientist and Citizen,* put it well: "Over periods of hundreds of years it is impossible to ensure that society will remain responsive to the problems created by the legacy of nuclear waste which we have left behind."[22]

"Legacy" is one way of stating it, but "curse" seems far more appropriate, for at the very least we are saddling our children with the perpetual custodianship of our atomic refuse. But it could be much, much worse than that, for we may be dooming them to any of a galaxy of horrors should the frail reservoirs we are building today release their virulent stew tomorrow.

We have little time left to reflect on our alternatives, for the time must soon come when no reversal will be possible. Dr. L. P. Hatch of Brookhaven National Laboratory vividly made his point when he told the Joint Committee on Atomic Energy:

> If we were to go on for 50 years in the atomic power industry, and find that we had reached an impasse, that we had been doing the wrong thing with the wastes and we would like to reconsider the disposal methods, it would be entirely too late, because the problem would exist and nothing could be done to change that fact for the next, say, 600 or a thousand years.[23]

To which might be added the following paralyzing thought stated by Dr. David Price of the U.S. Public Health Service:

> We all live under the haunting fear that something may corrupt the environment to the point where man joins the dinosaurs as an obsolete form of life. And what makes these thoughts all the more disturbing is the knowledge that our fate could perhaps be sealed twenty or more years before the development of symptoms.[24]

How fortunate, as professors Schubert and Lapp said, that our discovery of radiation is paralleled by our knowledge of its deleterious effects. Otherwise, ignorance of the consequences of irradiating man might induce great recklessness.

11

Thermal Pollution and Man's Dwindling Radiation Budget

Early in 1963, shortly after the nuclear power station at Indian Point, New York, began operation, an unusually large number of crows began to concentrate at a refuse dump near the plant. Curious sportsmen investigating the phenomenon were astounded to find that the crows had been attracted by dead striped bass—tens of thousands of them. "They were piled," according to one observer, "to a depth of several feet. They covered an area encompassing more than a city lot."[1]

What grim fate had befallen these fish? And who had carried them to the dump to rot as crow fodder?

Following their noses to a dock on the Hudson River, just below the power station, the sportsmen learned the answer to the first question. The dock was located at the shoreline where heated water, which had been used to cool the plant's atomic reactor, was being discharged back into the river. Beneath the dock was a ghastly scene: Countless dead and dying fish floated on their sides, their eyes bulging.

The stripers apparently had been attracted by the hot-water flow around the discharge pipe, then became trapped within the wharf and water intake structures at the plant. Further inquiry revealed that Consolidated Edison, which operated the Indian Point facility, was using two trucks to haul the dead fish to the dump.

The Great Indian Point Fish Kill, as that grisly episode has been called, is a nauseating but by no means atypical example of the devastating effects of what has come to be called thermal pollution. To recognize heat as a form of water pollution may require a bit of mental exercise, since we normally associate pollution with dirt and other visible contaminants. But if we define a pollutant as *any* factor whose introduction into the environment is detrimental or ruinous, then heat can surely be included in the growing list. For it was heat that drew those striped bass to the dock, and the hairsplitting explanation by one Con Edison official that the fish "were not *injured* by warm water —they were killed by the mechanical device"[2] in no way diminishes the gravity of the thermal pollution issue.

A rise in the temperature of a river can kill in many ways, some of them far more subtle than the outright slaughter that took place at Indian Point. The ecology of any given area—the relationship of each plant or animal to every other organism in the environment—is always an extremely complex and delicately balanced system. An apparently insignificant change in one branch of that system can have astonishingly significant effects on another, and indeed on interrelationships of *all* parts. The eradication of a large portion of a river's fish population, then, is not merely pathetic in terms of the fish and the humans who directly depend on them for food and sport; it may be tragic in the many unpredictably dire consequences suffered by all life in the environment.

Heat has a profound effect upon the composition of water itself. Warm water can hold less oxygen dissolved in solution than cooler water. Since fish must obtain needed oxygen from water by means of their gills, a drop in the oxygen content of water leads to suffocation and death. When a body of water already contains toxicants of other kinds near the maximum allowable tolerances—as is the case with an increasing number of America's waterways—even a relatively small rise in temperature can have disastrous consequences.

The collateral effects of water temperature changes are no less dangerous. Warmer waters produce much heavier growths of aquatic algae and other undesirable vegetation that turn clear and sparkling swimming areas into scum-ridden morasses. Of more immediate danger to wildlife, however, is that a rise of only a few degrees can enable formerly dormant fungi and parasitic organisms to flourish. Such has recently been observed in the Columbia River, essentially as a result of the discharges of the Hanford reactors. The net effect has been the growth of a deadly, but hitherto dormant, bacterial fish disease, *columnaris.* The disease has taken a heavy toll of salmon swimming upstream toward their spawning grounds.

Even if these fish are not slaughtered outright by excessive heat or *columnaris,* or even if excessive heat can be controlled, the fish population is nevertheless endangered by even moderate rises in water temperature. Cyclical changes in water temperature are vital to the reproductive processes of such fish as bass, trout, salmon, and walleyes. Unless they experience the natural rhythm of seasonal variations, their reproductive mechanisms will be disoriented, disrupted, or destroyed. Furthermore, a relatively slight upward change in water temperature can prove fatal to fish eggs and small fry, so that the attainment of the desired spawning grounds—an incredibly difficult struggle for many varieties of fish under the best of circumstances—does not guarantee the hatching of eggs or the development of the fry.

Just how great a problem is thermal pollution? Is it only a matter of killing salmon in Washington or stripers in New York? Scarcely, for *virtually every large fresh-water system in our country is earmarked for nuclear plant cooling purposes!*[3] A statement made by Dr. Gerald F. Tape, an AEC commissioner, at a 1968 Arkansas meeting of the Southern Interstate Nuclear Board, that "there is no basis for statements that 'thermal pollution from nuclear plants has been responsible for serious damage to marine ecology.'"[4] is therefore puzzling.

Of course, reactors are not the only thermal polluters; many other industrial processes, including conventionally fueled electricity production, spill high quantities of hot water into rivers and streams. However, nuclear-powered stations require up to 40 percent more water to cool their apparatus than do their fossil-fueled counterparts —as much as a half million gallons of water *per minute.*

Especially in southwestern states, this alarming high use of water has been a major factor in opposing nuclear plants. In areas where water is scarce, and where agricultural and residential users must always be aware of drought, the enormous water consumption of nuclear plants is not going to win many friends. Kern County, California, in fact, in a 1978 referrendum, defeated plans for a reactor complex in the town of Wasco.[5] The issue of water use contributed to the vote's results.

There *is* one way to combat thermal pollution, and that is by means of "cooling towers" designed to lower the temperature of the water emerging from reactors before it is returned to its source. Since these cost millions of dollars, however, utilities seek to avoid adding them to their facilities.

Scientists estimate that within a few years the plants producing the nation's electric power will require some 200 billion gallons of water per day, nearly all of it for cooling purposes.[6] As noted in the January–February 1968 issue of the *Sport Fishing Institute Bulletin:*

> This amount of water compares to an annual nationwide runoff totalling 1,200 billion gallons per day. In other words, a quantity of coolant equivalent to *one-sixth of the total amount of available fresh water will be necessary for cooling the steam-electric power-producing plants.* More ominously, during the two thirds of the year when flood flows are generally lacking, about *half the total fresh-water runoff will be required* for [steam-electric station] cooling purposes at inland locations. On certain heavily populated and industrialized northeastern U.S. watersheds, moreover, *100 percent of available flows may be passed through the various power-generating stations* within the watersheds during low-flow periods." [Emphasis added.]

And that takes us up only to 1980.

What has the Atomic Energy Commission and the NRC done to check the thermal destruction of our streams and estuaries? The AEC contended that the problem falls within the authority and jurisdiction of state agencies, so it is "unnecessary, and therefore undesirable, to involve the AEC." Some officials euphemistically termed the temperature effects of reactor discharges "thermal enrichment," and AEC Commissioner James T. Ramey told a congressional hearing in 1967 that "One point of view says it improves the fishing."[7]

The AEC made an attempt in the late 1960s to amend its rules to eliminate entirely the question of thermal pollution as irrelevant.[8] Fortunately, heightened environmental consciousness during the early 1970s forced the passage of legislation protecting the ecosystem. One of these laws, the National Environmental Policy Act of 1970, mandated that no major construction project can proceed without submitting an Environmental Impact Statement. This has reopened the question of thermal pollution (it has been a major issue in the Seabrook controversy) and has generally made utilities, and other corporate polluters as well, somewhat more responsive to the public interest.

Although the cost of building cooling towers is one no utility cares to pay, agreement to build one may make the public feel the utility is genuinely trying to meet the community halfway. Similar agreements may be made in matters of land beautification, such as building parks around nuclear facilities or offering to bury transmission cables underground. The community comes away feeling it has wrung a valuable concession out of the utility people, and it relaxes pressure or abandons the fight altogether. Meanwhile, attention has been distracted from a threat far more insidious than the overheating of water resources: the threat of radiological pollution of the environment.

At a colloquium held in December 1960 at the University of Chicago, Dr. Paul C. Tompkins of the U. S. Public Health Service pointed out that the quantity of radioactivity which our environment can tolerate is limited and cannot be used twice and that we therefore require a "budget" for environmental radiation—one that will allow not only for normal operations, but for accidents and engineering failures. Dr. Tompkins further stated that current radiation protection standards reflect only emissions from *particular* sources, not the accumulated radiation from *many* sources.

Far from heeding Dr. Tompkin's recommendation for a public health program, officials have ignored the concept of evolving a "radiation budget," in spite of numerous "routine" and accidental releases of radiation.

Consider, for example, just a few of the more recent accidental releases of radioactive material—bearing in mind that though most of these involve comparatively small quantities of radiation, it is the sum total which must concern us.

• **New Jersey:** "Officials at the Jersey Central Power and Light nuclear plant said Thursday that a small amount of untreated radioactive waste water was discharged into Oyster Creek early this week."
(*Atlantic City Press*, August 17, 1979)

• **Virginia:** "The North Anna nuclear plant had a history of small leaks of radioactivity and unplanned reactor shutdowns prior to this week's accidental release of contaminated gas, federal and utility officials revealed Thursday."
(UPI release, Sept. 28, 1979)

• **Arizona:** "An Army National Guard convoy carted 20 tons of radioactive tritium across central Arizona Saturday for burial at an ammunition depot, then technicians discovered a small leak in the leadlined steel box holding the gas.

"'Right now I can't answer what the potential hazard is,' Darrell Warren of the Arizona Atomic Energy Commission said . . ."

(AP report, Sept. 30, 1979)

• **Colorado:** "The Fort St. Vrain Nuclear Generating Plant was shut down Sunday because of a cooling malfunction and radioactivity was released into the atmosphere, officials reported."

(UPI release, Oct. 15, 1979)

• **Minnesota:** "Radioactive steam emissions from a ruptured tube at the Prairie Island nuclear plant apparently did not endanger the environment, although the steam spewed into the air for about 27 minutes, the Nuclear Regulatory Commission says."

(AP report, October 3, 1979)

Some radiation releases have involved considerable quantities of radiation. The incident reported in October, 1979 by *Environmental Action* is one example.

> What may well be the worst radioactive spill in U.S. history has gone largely ignored by the national news media. On July 16, a breach in a uranium mill tailings dam owned by United Nuclear Corp. in Church Rock, N.M. released 100 million gallons of contaminated liquids and 1,100 tons of solids into the Puerco River. . . .
>
> Samples taken after the spill revealed high concentrations of thorium, uranium, radium and lead in the water and along the banks of the Puerco 50 miles downstream.

Accidents aside, failure to officially measure and regulate routine emissions from the growing number of large nuclear power plants is, in itself, cause for serious concern. As an article in the *Saturday Review* for June 24, 1978 pointed out: "Daily operating emissions are measured—if at all—by the plants themselves . . . In repeated cases where environmentalists have insisted that government double-check certain plants, the agency has in fact found significantly higher releases than had been reported by the power plants." All of which should lead us to realize that the time is long past due to reconsider Dr. Tompkin's original proposal of working out a "radiation budget".

To fully understand the importance of this, a review of basic principles regarding radiation is in order.

In the first place, and foremost, many waste radionuclides take an extraordinary long time to decay. Cobalt 60 has a half-life of more than five years, strontium 90 of more than twenty-seven years, and cesium 137 of more than thirty years. Half-life is the time it will take for half of its atoms to disintegrate through fission. A few take far, far longer. A pound of carbon 14, could it have been placed in the tomb of Narmer, Egypt's first known pharaoh, about 3200 B.C., would still have more than half its potency today. A gram of plutonium 239 buried today will have lost only half its radioactivity in the year 25,969.

Thus, even though these long-lived isotopes may be widely dispersed in air or diluted in water, their radioactivity does not cease. It remains, and over a period of time it accumulates. It is therefore not pertinent to talk about the safety of any single release of radioactive effluents into the environment. *At issue, rather, is their duration and cumulative radioactivity.*

In addition, scientists do not understand all the ways in which radiation from specific isotopes affects human and other life. There are some two hundred isotopes created in the heart of a reactor, but analyses of their biological effects are scarcely complete, to say the least. The fact that the effects of atomic experiments and programs are often totally unexpected and are discovered only after radioactivity has been introduced in quantity into the environment was brought out in "Test Fallout and Water Pollution," an article by Dr. Barry Commoner in the December 1964 issue of *Scientist and Citizen.* Dr. Commoner said:

> Massive nuclear testing which began with the development of the hydrogen bomb in 1953 was well under way before most of its biological consequences were appreciated. The unanticipated tendency of world-wide fallout to deposit preferentially in the North Temperate Zone was unknown until 1956; the hazard from radioactive iodine and carbon-14 was not brought to light until 1957; the special ecological factors which amplify the fallout hazard in the Arctic were elucidated for the first time in 1960; experiments which suggest that strontium-90 may cause hereditary damage by becoming concentrated in the chromosomes were first reported in 1963."

A perfect example of this "information lag" may be seen in a story related by Dr. Commoner who is author of several important books on energy, including *The Poverty of Power.* Speaking before a conference on Nuclear Power and Environment held in Vermont in September 1968 (which the AEC declined to attend, incidentally), he told his audience of an important biological observation that, though several years old, seemingly had received little attention—in any event, he had just learned about it. It concerned the radioactive isotope iodine 131.[9]

Iodine 131 is one of the inevitable products of nuclear reactions, whether in bomb or reactor. It is relatively short-lived, with a half-life of only about eight days, so after a few weeks any iodine 131 released into the environment decays thoroughly.

Because of its short life, iodine 131 contamination from nuclear explosions was for a long time neglected. In time it was realized that actually it was among the most hazardous of all fission products, because iodine is concentrated in the thyroid gland, and even low concentrations of the radioactive variety would, in the process of decay, severely damage the thyroid and cause harmful biological changes among which is thyroid cancer.

One would think that in the absence of extensive nuclear weapons-testing or serious reactor accidents, the environment would be completely free of fast-decaying iodine 131. Dr. Commoner, however, had only recently been reminded of a test by investigators at the University of Nevada, who studied the iodine 131 content of cattle thyroids during a period (1959–61) in which there were only rare environmental intrusions of radioactive iodine from nuclear tests. "Much to their surprise," said Dr. Commoner, "—and I must confess, mine as well— they found, consistently, that even when there were no nuclear tests, cattle thyroids always contained *some* iodine 131."

What was most surprising was that the iodine 131 levels in the cattle thyroids are constant—month after month. Since iodine 131 decays so fast, it means that new iodine 131 is entering the thyroid as fast as the original iodine 131 decays. There must be a constant introduction of radioactive iodine into the environment, then—from a source other than nuclear explosions, which were ruled out because of their infrequency at this time.

After considering all possible sources, the University of Nevada investigators concluded, "The principal known source of iodine 131 that could contribute to this level is exhaust gases from nuclear reactors and associated fuel processing plants." Similar data, Dr. Commoner added, suggest that this situation is widespread, *involving at least the western third of the United States.*

Another example of the sketchiness of our knowledge of the effects of certain radioisotopes is tritium. Tritium is a radioactive isotope of hydrogen, which, if absorbed by the human body in place of stable, nonradioactive hydrogen, might have deleterious effects. Since it has a relatively long half-life, twelve and a half years, and its distribution into body tissues would be as wide as its parent element hydrogen, its presence in reactor effluents dictates the highest caution. Yet in his keynote address to the Health Physics Society Symposium at Atlanta, Georgia, on January 24, 1968, AEC Commissioner Wilfred E. Johnson admitted that the release into the atmosphere of tritium and noble gases such as the ten-year half-life krypton 85 would be a problem in the future, and that as yet scientists had not devised a way to solve it.

The other isotope Commissioner Johnson mentioned, krypton 85, gives us even greater cause for concern. This isotope, with a half-life of about ten years, is particularly difficult to extract from the effluvia of reactor operations. J. R. Coleman and R. Liberace of the Radiological Health Division of the U.S. Public Health Service stated in an official report dated November 1966 that removal of krypton 85 from reactor wastes involved "costly techniques which are reasonable only if very small volumes or flow rates are used"[10] — a condition that does not apply to the big reactors now built or presently to go into operation. And so instead of spending money to remove krypton 85, and thus raising the cost of "cheap electricity" to a realistic price that embraces health safeguards, reactor operators have no choice but to let the gas go into the air.

Although some radioactive elements seek particular tissues and organs when taken into the system, krypton 85, with its tendency to dissolve in fatty tissue, is distributed fairly evenly throughout the human body, raising over-all internal radiation and the attendant chances of cancer induction, genetic damage, and shortening of life.

By what amount will krypton 85 raise exposure levels? According to the highly detailed but conservative report made by Coleman and Liberace, the human whole-body dose from this isotope will be between twenty to one hundred millirads per year by 2060. We merely point out that the National Committee on Radiation Protection and

Measurement has set 170 millirads per year, above natural background radiation, as the maximum amount of radiation from any and all sources to which the general population should be exposed.

Thus the release of krypton 85 alone may exhaust about three-fifths of the human radiation "budget" within the next century!

It must here be pointed out that there is no safe level of radiation, and that all these figures—which many scientists have claimed are far too high—balance a risk against a supposed benefit. John Gofman and Arthur Tamplin have estimated that if everyone were to receive that 170-millirad dose, an additional 32,000 cancers would appear each year.[11] And it only takes one radioactive particle to act on one cell in order to kill or damage that cell. Higher exposures over larger areas of the body merely increase the chances of this undesirable event.

Considering, therefore, that krypton 85 is but one of a compendium of some two hundred radioactive elements produced by reactors, the reassurance that carefully monitored releases of low-level radioactivity into the environment are not harmful is nonsense.

Such reassurances are deceptive in another way. While, as we've seen, some radioactive elements taken into the body are distributed evenly, others are absorbed into specific tissues and tend to build up concentrations. Iodine 131, for instance, seeks the thyroid gland; strontium 90 collects in the bones; cesium 137 tends to accumulate in muscle. Many of these isotopes have long half lives, some measurable in decades. Human intake of these poisonous isotopes does not, therefore, lead merely to general distribution throughout the body. It leads instead to build-ups in the specific tissues or organs to which those isotopes are attracted, thus increasing by many times the exposure dosage in those local areas of the body.

Man is by no means the only creature in whom radioactive isotopes may concentrate. Indeed, the dietary needs of all vegetable and animal life dictate the intake of specific elements. Those elements, whether radioactive or not, will concentrate even in the lowest and most basic forms of life. They are then passed up food chains, chains such as grass-to-cattle-to-milk-to-man. As they progress up these chains, the concentrations increase, sometimes by hundreds of thousands of times. Norman Lansdell, in *The Atom and the Energy Revolution*, reports a study of the Columbia River in the western United States in which it was found that while the radioactivity of the water was relatively insignificant, (1) the radioactivity of the river plankton was 2,000 times greater; (2) the radioactivity of the fish and ducks feeding on the plankton was 15,000 and 40,000 times greater, respectively; (3) the radioactivity of young swallows fed on insects caught by their parents in the river was 500,000 times greater; and (4) the radioactivity of the egg yolks of water birds was more than a million times greater.[12] This is the same process by which other poisons—pesticides and the like—concentrate in deadly doses, although they are extremely dilute at the source.

A closer look at this amplification process is enlightening. For example, zinc 65 is produced in a reactor when chain-reaction neutrons interact with zinc in reactor components.[13] Some of this radioactive isotope is released into the waters surrounding reactors. Studies of numerous sites downriver from reactors, such as the Columbia River on which the Hanford, Washington, reactor is located, or Connecticut's Thames, below the Groton nuclear submarine station, have disclosed concentrations of zinc 65 (among other isotopes) in algae, fish, oysters, vegetation, and animals that feed or prey on those foods.

Scrutiny of the wildlife in a pond receiving runoff from the Savannah River plant near Aiken, South Carolina, demonstrated that while the water in that pond contained only twenty-five thousandths of a picocurie per gram of zinc 65, the algae that lived on the water had a concentration of 148 picocuries per gram, an increase of 5,820 times. The bones of bluegills, an omnivorous fish that fed both on the algae and on fish that ate the algae, had 218 picocuries per gram, a concentration of 8,270 times the amount of zinc 65 found in the water. Another example: In 1958, when Columbia River water had about 188 picocuries per kilogram, oyster samples at the mouth of the river had concentrations as high as 63,500 picocuries per kilogram, a 330-fold concentration.

Studies of humans in the Columbia River area drive the point home. Measurement of radioactive zinc in the bodies of people who drink Columbia River water showed that these people have 57 picocuries of zinc 65 per kilogram of body weight—more than 4,000 picocuries in a 154-pound man. A man drinking three glasses of milk and eating about a quarter of a pound of meat daily from the Columbia River Valley would have nine times that amount of radiation in his system—which is greater than that measured in some persons working in atomic energy installations.

Here then are clear illustrations of the way in which almost undetectable traces of radioactivity in air, water, and land may be progressively concentrated so that by the time they end up on man's plate or in his glass they have become a tidy package of poison. This is to say nothing, by the way, of genetic effects and bizarre transformations, about which practically nothing is known.

From 1946 to 1962, 47,500 fifty-five gallon drums were thrown into the Pacific Ocean off San Francisco.[14] The Environmental Protection Administration estimates that as many as one-fourth of these drums have failed, leaking low-level radiation into waters rich in edible fish. Giant sponges, some three feet long, have apparently attached themselves to some of these barrels.[15]

That *low-level* waste is, in the light of these discussions, a grossly deceptive term is obvious. The level is low only if one thinks of the dispersion of radioactive particles into air or water on a single occasion. If one considers instead the accumulation of long-lived isotopes released over decades by scores of reactors and other sources, and if one considers the reassembly of those scattered particles in growing concentrations up the stages of the food chain, and if one considers their further concentration in bones, muscles, glands, and other tissues and organs of the human body—where, then, is the "low level"?

This gross misconception has led to formulation of a very dangerous and deceptive system of radiation guidelines established by various groups and adopted by our government. Altogether, these guidelines lay down a "maximum permissible dose" that an individual can receive from any single radioactive source without suffering harmful effects.

The fallacy in the "maximum permissible dose" system is that the guidelines fail to take into account the *total* accumulation of radiation we receive from *all* emitting sources. Ritchie Calder expressed it clearly when he said: "Radiation adds up to a sum. For example, when a series of operations feeds legalized waste into a common stream, the contribution of each may be insignificant but the total may be really significant. Our common environment is like that stream."[16]

The maximum permissible levels of radiation exposure used in the AEC's regulations are based on the recommendations of the National Council (originally Committee) on Radiation Protection and Measurements—the NCRP. This private organization, founded in 1929, has been the principal standard-setter for radiation protection in the United States. But although it is known as a body of distinguished and cautious scientists and engineers who try to perform their functions conscientiously and objectively, a flaw in its fundamental approach is not widely recognized. That flaw is described by Harold P. Green in an article in the November 1967 *Bulletin of the Atomic Scientists:*

> Despite the competence, conservatism, and integrity of the NCRP, its role raises significant questions of public policy. Its public pronouncements explicitly reflect the policy of "calculated risk"; in making its recommendations, it balances the benefits of radiation technology against its hazards. It endeavors to adopt levels of permissible exposure which will protect the public but which will not, at the same time, stifle the radiation industry. As Dr. Lauriston Taylor, for many years the head of NCRP, has pointed out, NCRP and its international counterpart are ". . . scientific and technical in their makeup and hence do not attempt to solve the social problems that radiation control may introduce," and ". . . the establishment of permissible levels of radiation exposure is not basically a scientific problem . . . it is more a matter of philosophy, of morality, and of sheer wisdom." Despite these disclaimers of competence the NCRP has, largely by default of others, undertaken to make recommendations as to "maximum permissible dose."

Green's conclusion is of the highest importance to our discussions:

> It is apparent that NCRP, in adopting radiation protection standards which are arrived at through the balancing of social values, and which are more or less automatically incorporated into government regulations and used as standards by the courts, is performing a legislative function. As such it is unrepresentative, since it consists only of scientists and engineers, without membership from other disciplines (law, psychiatry, the clergy, economists, etc.) and societal interests. Nor does NCRP have political accountability of the kind that should be inherent in lawmaking bodies.

The fallacy of the "maximum permissible dose" can lead society far astray from the path of safety and health. "A case in point," states Walter Schneir in an article in *The Reporter,* "is the pollution of the Animas River in Colorado and New Mexico, where the water used by thirty thousand people was found by the Public Health Service to contain radium far in excess of maximum permissible levels. For a proper assessment of the danger from the wastes, the Public Health Service sought to learn what other man-made radiation the people of the area were exposed to. By unhappy coincidence, the peas, cabbages, lettuce, and other vegetables grown in this area were also discovered to have extremely large amounts of strontium-90 from fallout."[17]

The National Committee on Radiation Protection and Measurements, in an article in *Science*, advocated that "the levels should be set so that the typical person in the area will not receive more than the established permissible dose when all sources are combined." And Dr. Theodore Rice, laboratory director of the U.S. Bureau of Marine Fisheries, Radio-Biological Lab, Pivers Island, Beaufort, North Carolina, was reported as stating that at present the AEC Code of Federal Regulations, Title 10, Part 20, on maximum permissible limits of release of radioactive wastes, was so flexible as to be no regulation at all. "Although safe levels of radiation are determined on the basis of intake of drinking water, if we get additional amounts from food and other sources, we get more than the maximum level."[18]

While more traditional conservation issues such as thermal pollution have in the past obscured the far more serious issue of radiation —the successful fight to stop a proposed plant on Cayuga Lake in New York being a case in point[19]—this is no longer the case. In recent years a heightened public consciousness about the effects of radiation has resulted in considerable turmoil within and without the atomic establishment. And in the next few years, after the terror of Three Mile Island, we can expect that many plants will be dealt sharp blows over the issue of radioactivity containment. John Gofman had pointed out that even if the atomic industry is able to contain 99 percent of its wastes (most unlikely, if past history is any indication), the hundreds of pounds released could result in as many as 500,000 more lung cancer deaths per year, if the nuclear economy reached the proportions once proposed.[20]

Gofman further poses this query: "Many serious public health experts consider 63,500 lung cancer fatalities per year [the present amount] to represent a most serious epidemic. How should they view the burgeoning plutonium based nuclear fission energy economy, proceeding under regulatory standards that would permit a four-fold increase supplementary to this epidemic?"[21]

Our seas are grand, but they are not of infinite capacity. They can accommodate a huge amount of waste and flotsam, but, as we are discovering with all natural systems on our planet, they have limits. It might take a good deal of radioactivity to contaminate the seas, but the enormous potency of nuclear material, its longevity and cumulative lethality, could in time raise radiation levels by a small but undesirable, if not alarming, amount. Runoff of radioactive efflu- ents and bomb tests or peacetime nuclear excavation, the sinking of nuclear ships or of ships and planes bearing nuclear material, large- scale releases resulting from accidents on land- or island-based reac- tors, and of course the continued ocean-dumping of enormous quanti- ties of atomic waste—these can all make their contributions.

In recent years scientists have been looking with increased hope to the sea as a source of food and other resources. Under the Pacific Ocean alone, it has been estimated, there is enough aluminum to last at the present world rate of consumption for *eighty thousand* years, compared to a hundred years of reserves known on land; enough manganese for *four hundred thousand years*, compared to land reserves of a hundred years; and enough copper for *six thousand years*, compared to land reserves of only forty years. Mismanagement of the oceans and sea beds, however, may render it more difficult, if not impossible, to harvest this incredible bounty.

This is seen most clearly in the contamination of plankton, the minute plant and animal organisms that float on or near the surface of the sea. An article in *The Nation* found the plankton at one site to be 150,000 times more radioactive than the water. But while the con- tamination of such microscopic lowlife would not, at first glance, seem to have much bearing on the fate of our planet, the upshot could range from the "merely" catastrophic to the totally apoca- lyptic.[22]

It has long been known that protein-rich plankton is one of the prime sources of food for fish. Radioactive poisoning of this food, therefore, could eradicate it as completely as a flood eradicates a farm crop, leading to starvation of whole species of sea life and tur- bulent changes in marine ecologies. Even if the plankton were not actually poisoned out of existence, its ingestion by fish, and the con- centration of radioisotopes in their bones, tissues, and organs, could lead to wholesale extermination or deadly mutations eventuating in extermination. In either case the life of the sea, and the life that depends on it, would be radically affected by radiological contamination of this elemental life form.

But that is not all. Efforts are now under way to harvest plankton as a potentially vast source of protein for a humanity that is growing by one million persons per week. Even if nothing is said about loss of edible fish to world markets as a result of radioactive contamination, the loss of edible plankton would strike a grave blow to a major hope for the relief of future famines.

Yet even this prospect is eclipsed by a potential hazard of awesome proportions—one which threatens to transform the very air we breathe.

One man who has considered these possibilities is Dr. LaMont Cole, professor of ecology at Cornell University. Most of us who got through high school biology may still be familiar with the cycle by which oxygen and nitrogen are kept in immutable balance in the atmosphere. By photosynthesis green plants produce oxygen with the help of sunlight. Nitrogen is produced and released in the decay of dead plants and other organic life. Plants and animals need both elements to form their proteins.

Professor Cole has pointed out that more than 70 percent of the world's photosynthetic oxygen is produced by marine microorganisms floating on or near ocean surfaces. Introduction of biologically active materials such as radioisotopes, to which the ocean's living forms have never before had to adapt, could prove calamitous. "No more than a minute fraction of these substances or the combinations of them have been tested for toxicity to life," says Dr. Cole, "—to the diatoms, the microscopic marine plants that produce most of the earth's oxygen, or to the bacteria and micro-organisms involved in the nitrogen cycle."[23]

Considering all we already know about the effects of radiation and all that we do not yet know, (and may learn too late), the American people are entitled to a careful, comprehensive study to determine how much radioactivity has already been introduced into the environment. Account should be taken in this study of all atomic reactors now in operation; the radioactive wastes they produce; deliberately induced reactor accidents aimed at determining effectiveness of safety devices; and various experiments such as Plowshare. All other nuclear accidents to date, the disposal of radioactive wastes in rivers and oceans, and radioactivity resulting from weapons-tests should also be included. Furthermore, current concentrations of radioactivity in flora and fauna along various ecological chains should be measured, so that changes could be monitored.

Only in this way can the American people know exactly how our radiation budget stands at the present moment.

12

Absolute Power

The Atomic Energy Commission woke up in the mid-1960s to discover that the robot it had for twenty years attempted to breathe life into had risen from the table and was marching on the nation's cities. Almost all of the hazards to which this nation is now subject owe to the fact that the regulatory forces which ordinarily would have contained the monster had been subdued, permitted to atrophy, or actually has been enlisted into the service of atomic power's advocates. A detailed examination of this process will place those perils in the context of the policies that created them.

Four areas stand out distinctly in this connection: (1) the internal components of the Atomic Energy Commission itself, such as advisory committees and review boards; (2) the congressional committee charged with review of the AEC's activities; (3) the state and local governments; and (4) the public at large.

The first area in which the AEC's near omnipotence was manifest is in its relationships with its own advisory groups. The Atomic Energy Act of 1954 authorized the AEC to establish subcommittees to conduct administrative proceedings, public hearings, safety examinations, licensing reviews, and so on. Several of these groups are implicitly designed to provide conservative drag on any tendencies to accelerate the commercial reactor program hazardously.

In practice, however, that conservatism has been all but nonexistent. The AEC—and later the NRC—was not obliged to confer final jurisdiction on subcommittees or to accept their recommendations. The commission is entitled to disregard or overrule adverse findings, and it has even, as we have seen, sought to suppress them. The statutes make it possible for the commission to bestow diluted responsibilities on subcommittees so that their reports often constitute a virtual playback of the commission's positions.

Such is not uncommon practice in a hierarchical system of administration. However, in an area as supremely delicate as atomic power, advisory and review groups must have the widest possible latitude in determining the suitability—or, more to the point, the *unsuitability*—of plant blueprints, reactor safeguards, sites, radiological security, inspection procedures, and the like. Adverse findings in these areas made by a subcommittee must be given profound weight when the commission considers issuance of a license; in fact, such adverse findings should, one would think, be tantamount to an automatic red light. One could go even further to suggest that even if a subcommittee gives its *approval* to an application, the minority report of dissenting members should cause the parent commission seriously to hesitate, though the minority report be filed by but one member.

Although it had been apparent for years that the combination of promotion and regulation of atomic energy in the AEC had caused the promotion angle to be encouraged, to the detriment of the safety issue, the AEC tenaciously resisted attempts to separate its two functions, claiming that communication between the two bodies would be difficult. In 1974, however, the AEC, which had already been weakened by such factors as the National Environmental Policy Act of 1970, requiring environmental impact statements and thus opening up hearings to include questions the AEC would have preferred to ignore, was successfully fissioned.

Effective January 1, 1975, the Energy Reorganization Act of 1974 broke the AEC into two bodies: the Nuclear Regulatory Commission, charged with protecting the public, and the Energy Research and Development Administration (ERDA), designed to promote atomic and, in theory at least, other types of power production. The changes, however, were largely cosmetic. Of ERDA's employees, exclusive of outside contractors, who were also nuclear-dominated, 84.3 percent were hired out of the old AEC.[1]

Therefore, it is not surprising that ERDA's 1977 budget—its last before being absorbed into the newly created Department of Energy—requested more than 62 percent of its allocation (up a percentage point from the previous year) for nuclear fission and fusion. The remaining 37.1 percent was scattered among fossil fuels, solar, geothermal, conservation, and other sources.[2] About one-third of ERDA's budget, incidentally, is earmarked for military research; in other words, nuclear bombs and other deadly misuses of technology.[3]

The "revolving-door" factor—the switching of personnel between regulator and regulated—has of course been in operation in both ERDA (now the Department of Energy) and the NRC. In the latter case, according to the citizen lobby Common Cause, 279 of the top 429 NRC employees "came from private enterprises holding licenses, permits, or contracts from NRC. This represents 65 percent of NRC's top personnel."[4]

The door swings the other way too. Former AEC commissioners, such as James T. Ramey, former members of the congressional Joint Committee on Atomic Energy (JCAE), such as Craig Hosmer, and former high executives within the AEC, such as Robert Hollingsworth, have all accepted lucrative positions within the nuclear industry after leaving the government.[5] And the NRC has been criticized for many of the same abuses formerly attributed to the AEC. One important difference now, though, is that criticism is increasingly coming from the NRC's own staff.

In February 1976, for instance, shortly after engineers Dale Bridenbaugh, Richard Hubbard, and Gregory Minor resigned from lucrative positions within General Electric "to commit ourselves totally to the education of the public on all aspects and dangers as we have learned them over our many years of experience in the industry,"[6] Robert Pollard, NRC's project manager for Indian Point #2, submitted his resignation. Calling the Indian Point plant "an accident waiting to happen," Pollard noted that Indian Point #2 was by no means the only such reactor. He stated further, in testimony before the Joint Committee on Atomic Energy:

As a result of my work at the Commission, I believe that the separation of the Atomic Energy Commission into two agencies has not resolved the conflict between promotion and regulation of commercial nuclear power plants. Because I found that the pressures to maintain schedules and to defer resolution of known safety problems frequently prevailed over reactor safety, I decided to resign. I could no longer, in conscience, participate in a process which so effectively evades the single legislative mandate given to the NRC—protection of the public health and safety.[7]

NRC management officials are aware of many unresolved safety problems that apply to whole classes of reactors, but they continue to press for the expedition of individual plant licenses. This NRC policy is so absurd that it bears repeating: It seems that unresolved problems that are relevant to many reactors need not be resolved and should not be discussed in deciding whether to license a particular reactor.[8]

Some months later, another NRC employee, this time a safety analyst named Ronald M. Fluegge, also resigned.

I have been repeatedly frustrated in my effort to make the agency deal honestly with pressing nuclear safety problems. . . .

I believe—in common with a substantial number of my colleagues on the NRC technical staff—that NRC is violating its public trust. . . .

Time and time again, the NRC has covered up and brushed aside nuclear safety problems of far-reaching significance. We are allowing dozens of large nuclear plants to operate in populated areas despite known safety deficiencies that could result in very damaging accidents. We are issuing Safety Evaluation Reports that are carefully censored to conceal major safety problems. We are withholding from the public NRC staff technical analyses of a wide range of unpleasant nuclear safety difficulties. We are giving the public glib reassurances about nuclear plant safety that we know lack an adequate technical basis. NRC management has let its commitment to the industry . . . compromise the agency's regulatory integrity.

Two and a half years before the Three Mile Island pressurized water reactor accident, Fluegge went on to say:

I certainly share his [Pollard's] concerns about Indian Point-2, which surely would be shut down by any prudent nuclear safety regulator, and have indeed advised the NRC management of safety problems that require the immediate shutdown of Indian Point-2 *and of all presently operating commercial PWR nuclear plants in the U.S.*[9]

Fluegge had been afraid to speak out until he had found a job outside of the nuclear establishment, but he is on the record *in favor of* nuclear power. Unlike most proponents, however, he insists on attention to reactor safety.

When a utility application is filed for construction of a nuclear plant, it is accompanied by the company's analysis of the safety features to be incorporated. This analysis is submitted to its own regulatory staff and the Advisory Committee on Reactor Safeguards. Both groups review the safety features and file their findings with a three-man Atomic Safety and Licensing Board, consisting of a lawyer and two technically qualified persons drawn from a twenty-six-member panel established by the NRC. The latter holds a hearing based mainly on documentary evidence, and unless the case is contested, or one of the two review groups (or both of them) report adversely on the application, the board will give the go-ahead to the applicant.

The system sounds well safeguarded against manipulation by overeager proponents of nuclear power, but close examination reveals many discrepancies.

For one thing, disputes between these two groups are kept most strictly "in the family." In January 1965 the AEC appointed a seven-man regulatory review panel "from outside the Government" to review the commission's policies and procedures for licensing nuclear reactors. The chairman was formerly the AEC's general counsel; the other six were scientists, engineers, and executives drawn from the nuclear industry.[10] It is not, then, surprising to learn that one of the panel's recommendations was that every effort be made to conceal from the public any differences of opinion between the ACRS and the AEC's regulatory staff.

But there is far more serious evidence that the commission's safety review groups have little independence. Approval by the regulatory staff of an application has, it has been asserted, become more and more predictable. As Harold P. Green, professor of law and director of the law, science, and technology program of George Washington University National Law Center, points out in describing the regulatory staff safety report, "The report characteristically concludes with a judgment that safety standards are met and a recommendation that the construction permit be issued."[11]

The Advisory Committee on Reactor Safeguards seems slightly more independent than the regulatory staff, but its rubber-stamp okay of virtually every application is ensured by the concept of the "provisional construction permit." Before describing it, a little background will be helpful.

The ACRS is a prestigious group of technical experts, which, though its members are appointed by the NRC, has tried to maintain a critical, conservative, and independent posture. Unfortunately, over the years the ACRS has had a steadily decreasing influence over the NRC. And yet, so important does Congress regard the ACRS's function that in 1957 it passed legislation placing the committee under the legislature's aegis, making it mandatory for it to report on reactor safety, and making the report part of the public record of proceedings. The idea behind this extraordinary move by Congress, obviously, was to create a "conscience" for the AEC, and to make sure the parent organization did not with impunity dismiss the urgings of the conscience. Congress knew what it was doing, for not long after the Atomic Energy Act of 1954 went into effect, the AEC found itself suffering from rather distressing pangs of "conscience."

The struggle reached a climax in 1956 with the review of the construction license application for the Fermi plant. In that case, as we saw in chapter 1, the ACRS reported to the AEC that "the Committee believes there is insufficient information available at this time to give assurance that the PRDC reactor can be operated at this site without public hazard." Pressed heavily by the applicant, committed to a timetable, and, above all, fearful that the negative report might spook everybody preparing to put money in the atomic power program, the AEC not only overruled the ACRS, but, to use Representative Holifield's word, "suppressed" the report. The voice of expediency was clearly evident in the AEC's ruling:

> We believe the public interest in the development of the fast breeder reactor, the time to be saved in proceeding with construction while the remaining technical and safety problems are being solved, and our responsibilities under the Atomic Energy Act of 1954, are better served by continuing the permit.[12]

Although we have recorded the angry reaction of Senator Clinton Anderson, then chairman of the congressional Joint Committee on Atomic Energy, when he learned of this action, it is worth repeating for its relevance to the issues in this chapter:

> The issuance of this construction permit, in my opinion, sets a dangerous

pattern in the early stages of AEC regulative and quasi-judicial activity. . . . From a practical standpoint, AEC might feel obligated to go through with a bad deal with respect to public safety because they will have permitted the expenditure of huge sums under the construction permit. It is my belief that decisions on safety should be made without any examination of dollars involved, but only from the standpoint of human lives.

Senator Anderson, of course, helped organize the opposition to Fermi. But the fact that the Supreme Court upheld the Fermi applicants strengthened immeasurably the AEC's power—even though the nearly catastrophic failure of the reactor vindicated all of the ACRS's worries.

Most important of all, the AEC's triumph in the Fermi case established the concept of the provisional construction permit. This permit acknowledges that the applicant's safety proposals are not adequate, or the technology for certain proposed safety features is not fully perfected, but nevertheless permits the applicant to proceed with construction *in the expectation that the problems will be resolved by the time the plant is up and ready to apply for an operating license.*

This provisional scheme serves everybody's purpose very nicely. It enables the applicant to go through with construction without having to resolve important safety problems first, or without having to wait for technology to resolve them for him. It does not commit the NRC to approve all proposals for safety components until the plant is finished; and, theoretically at least, it forces the builder to proceed at his own risk, because there is no guarantee that approval will be automatic when the plant is finished. Of course, in practice, "momentum is on the side of the applicant, not on the side of the public," as the minority report in the Supreme Court's Fermi decision expressed it. The utility, having a huge investment to protect, and itself under pressure from the community to get the plant in operation before a power shortage occurs, is undoubtedly going to push very hard for final approval.

Finally, the provisional setup saves the face of the ACRS. It enables the committee to raise the alarm on inadequate safeguards just as loudly as it cares to, yet by granting conditional approval anyway, the committee manages to avoid running athwart of the NRC.

The provisional permit scheme stands close to center in the galaxy of reckless procedures described in this book. In *The Technology of Nuclear Reactor Safety*, T. J. Thompson and J. G. Beckerley clearly point out that "the important economic and safety decisions are made early in the reactor design stage. There may be no retreat from poorly made initial decisions. It is therefore essential that considerations of safety are given a vital role at this stage of a reactor project."[13] The book from which that passage is quoted was published under the auspices of the AEC.

The commission, of course, publicly proclaims its support of this principle. In answer to a question posed by the Joint Commission on Atomic Energy in its 1967 hearings, for example, AEC Chairman Glenn Seaborg stated:

> From the regulatory standpoint, it is important that a comprehensive safety review be conducted before large financial commitments have been made, designs frozen or construction accomplished. From the industry's point of view, it is equally important that this review be accomplished before irrevocable commitments have been made.[14]

But while the necessity for correct initial decisions on reactor design is acknowledged at congressional hearings, the *practice* of insisting on safe design before construction begins is obliterated by the provisional construction permit. On Tuesday, May 28, 1968, for instance, the ACRS reported to the AEC on five proposed nuclear power plant projects it had reviewed.[15] Though the layman may not be qualified to judge the technical merits of the committee's reports, it is by no means beyond his grasp that extremely serious dangers were ascertained by the committee. Here are excerpts from four of the five reports:

> The Committee believes that the proposed off-site power system should be modified to fulfill Criterion 39 so that no single failure will prevent the operation of minimum electrically powered safety features necessary to protect the core. (Crystal River Unit #3, Florida)

> The applicant is performing a detailed analysis of the effects of sudden failure (e.g., by seizure) of a main coolant pump. . . . The Committee continues to believe that control and protection instrumentation should be separated to the fullest extent practical. There remain questions in this area on the Point Beach design. (Point Beach Nuclear Plant, Unit #2, Wisconsin)

> The Committee continues to believe that control and protection instrumentation should be separated to the fullest extent practical. There remain questions in this area on the Kewaunee design. (Kewaunee Nuclear Plant, Wisconsin)

> The emergency power system originally provided for Units 1 and 2 has been

redesigned and expanded to serve all three units. . . . The design as proposed appears marginally acceptable. Questions arise regarding the capacity of the diesel-generators and regarding the necessity for paralleling of generators at some time after an accident. Consideration should be given to improvement of the system. (Browns Ferry Nuclear Power Station, Unit #3, Alabama)

The latter is doubly significant because it is the third unit of the stupendous TVA project which ACRS member Stephen H. Hanauer found so hazardous that he filed a public dissent, a precedent-shattering move. Not only was the AEC going ahead with the first two units, which the ACRS had found worrisome enough, but it was considering a third unit before the first two were even built, let alone proven.

The fifth of the five reports is particularly noteworthy for the reservations expressed by committee members, two of whom saw fit to add separate statements to the main text of the document. These statements are, at the risk of trying the reader with material of a semi-technical nature, presented here in their entirety. The first was written by Dr. David Okrent, who in 1966 was chairman of the ACRS:

The Fort St. Vrain Station [in Colorado] will have the first prestressed concrete reactor vessel designed and constructed in the United States, although such vessels have been built abroad. However, even abroad, only limited experience exists with these vessels. Not all of that experience has been favorable, and none of the existing experience covers more than a fraction of the operational life of the vessels. Only a limited amount of safety research work has been done in connection with various failure modes of these vessels, or on the effects of anomalies and errors of design, construction or operation. History teaches us that errors and misjudgments have been and will be made in the design and construction of vital components. The chance of such errors is increased when a long experience with design, construction, and operation is not available. At this time, it is not clear to me that significant faults in a prestressed concrete reactor vessel would necessarily be detected prior to the loss of integrity of the vessel. The inaccessibility of the vessel liner, cooling tubes, and thermal insulation compound this difficulty.

From the standpoint of reactor safety, the Fort St. Vrain design is especially vulnerable to vessel failure because a single structure serves as both reactor vessel and secondary containment. I believe it acceptable to construct the Fort St. Vrain station, in view of the remote character of the site, the moderate power of the reactor, the apparent great conservatism in the design of the reactor vessel, and the fact that only one unit is involved. However, I believe that it would not be prudent at this time to construct larger reactors of the Fort St. Vrain type at more populated sites without additional safety features to cope with major accidents involving various modes of failure of the reactor vessel.

The second statement was made by Dr. Joseph Hendrie:

I believe the Fort St. Vrain reactor should be contained in a building of such design pressure and leakage characteristics as to protect the public in the event of a major failure of the reactor vessel. I do not agree with the applicant's argument that the present design of the reactor vessel provides both primary and secondary containment of the reactor in an adequate manner. The great merit of the traditional secondary containment building is that it is a separate and independent barrier to protect the public from the effects of failures of the primary system. In the Fort St. Vrain design, this essential separation is lost, and the safety of the public depends upon the integrity of a single structure. The applicant concludes that a significant loss of integrity of the reactor vessel is impossible, due to the reinforced, prestressed concrete construction. This may be a correct conclusion, but in a matter as important as the public safety I believe it should be supported by a substantial amount of favorable experience in the construction and operation of high-temperature, gas-cooled reactors with concrete vessels. In the absence of such experience, I believe the Fort St. Vrain reactor, and any similar units that might be proposed in the near future, should have secondary containment buildings.

But despite the obvious hazardousness of these five nuclear plant proposals, and despite the undeveloped technology in many critical areas—in short, despite every good reason to withhold approval of a license until all crucial problems were solved—the Advisory Committee in every single case ended its report with the following words, or slight variations:

The Committee believes that the various items mentioned can be resolved during construction and that the proposed power plant can be constructed at the site with reasonable assurance that it can be operated without undue risk to the health and safety of the public. [Emphasis added.]

So far, of all the plants constructed under this scheme, only one —the twin reactor at Diablo Canyon, California—has not obtained an operating license since construction. Even the NRC has been unable to deny compelling evidence of a major earthquake fault, the Hosgri, running almost directly under the plant. The fault may have been responsible for a devastating quake in 1927.[16]

Six years after construction, the NRC has neither issued nor flatly refused to issue the operating permit; the plant is a billion-dollar white elephant. The NRC had reportedly been ready to allow the plant — considerably beefed up — to operate, when the accident at Three Mile Island changed plans.[17]

The Atomic Energy Commission made other moves to make sure its safety review process moved along at a vigorous clip without annoying delays. For instance, the AEC once proposed legislative changes that would relieve the ACRS of the mandatory case-by-case review requirement. The reasoning behind this change was that a number of proposed reactors were similar to types already approved. In some cases a utility will apply for a second or third unit identical to one already built or approved: Why make the ACRS go through review procedures in such cases?

The answers, of course, are obvious. First, research and development might have come up with safeguard improvements since the first reactor was built. Second, the addition of a second, third, or fourth reactor on the same site demands a new look at cooling water resources, emergency electrical systems, interconnection of safeguards, siting, availability and training of personnel, structure of control systems, and so forth. Third, the building of a reactor in, say, Wisconsin that is identical to one in New Jersey may still present a host of new siting considerations demanding modification of plans. Fourth, the *original* reactor may, upon going into operation, have proved faulty or hazardous, so that "copies" of it will duplicate the same unsafe features.

More recently, President Carter, under the persistent urging of James Schlesinger, who recently resigned as Secretary of Energy — and who once chaired the Atomic Energy Commission — has made a great deal of noise about speeding up and streamlining licensing procedures. Citizen involvement, under the Carter plan, would be all but eliminated. Carter has reaffirmed his interest in such a plan since the Three Mile Island affair. This is all the more ironic, as Carter was elected in 1976 claiming nuclear power would only be encouraged as a "last resort."

Most important of all, reactors are not assembly-line products about which it can be assumed that if one is good the rest are just as good. In the 1967 congressional hearings, Harold L. Price, AEC's Director of Regulation, stated: "We just haven't seen any standard—what you would really call a standard—reactor yet." Then in the next breath he said:

> The only duplicates we have seen are the twins where a few of these companies have applied for two identical plants to be built at practically the same time on the same location.
> If I may be a little more facetious, we handle that second one real fast.[18]

Perhaps it was remarks like these that Justices Hugo Black and William Douglas anticipated when they referred, in the Fermi case, to "a light-hearted approach to the most awesome, the most deadly, the most dangerous process that man has ever conceived."

And, in fact, until 1971, all reactors were licensed as experiments.

The Atomic Energy Commission also managed in great measure to neutralize the congressional committee charged with the stern duty of overseeing the AEC's activities, the Joint Committee on Atomic Energy.

Admittedly the Joint Committee was not so much a regulatory body as a review panel serving to keep Congress informed of atomic energy activities, to keep the AEC informed of congressional moods and measures, and to act as go-between on atomic energy legislation. However, being an arm of the legislature, the Joint Committee could if it chose wield powerful influence over the AEC, for implicit in the legislative review system is the threat of sanctions imposed by Congress if it is not satisfied with what the committee reports. Even if that were not the intention, the joint committee is constantly in a position to assume such powers at its discretion.

Early in nuclear power history, the committee did play this "watch-dog" role, its hearings being conducted with a certain arm's-length formality with the AEC, suggesting that a highly critical eye was being cast on the AEC's activities. Still, many of the committee members were, by admission or deed, most sympathetic to the cause of atomic power, and when the great atomic power promotion got under way, these members were in the vanguard.

A crisis occurred in the Fermi case, when some committee members, including the chairman, cried "Foul" over the suppression of the ACRS's negative safety report and encouraged legal action to stop the building of Fermi. The Supreme Court's decision severely undercut the congressional committee's power, however.

Reading the transcripts of the 1967 joint committee hearings on atomic power, one is puzzled by the seeming lack of skepticism, anxiety, or dissatisfaction expressed by the legislators about the AEC's and the industry's activities. There are few conflicts or dramatic clashes. Sometimes the reader's pulse quickens as the questioning turns up an alleyway bristling with the makings of a good fight. But alas, the interrogators step away, declining to engage. Typical is the following exchange between a committee member and AEC Chairman Seaborg:

REP. HOSMER: I have only one question. In years past we used to hear the argument that while the Commission had this licensing and safety responsibility, and at the same time it had a promotional responsibility in connection with the nuclear industry, that somehow there was a conflict between the two—that it might be detrimental to the full and complete discharge of the AEC's licensing and safety responsibilities.

You have recited the magnificent safety record we have. I don't hear this talk much anymore but do you have any feelings about whether there is a conflict or not?

DR. SEABORG: As I have watched it, and with the separation of the regulatory function as we have it now, a step that we took in 1961, the Commissioners are pretty careful and quite meticulous in their regulatory responsibility and have this possibility of conflict in mind. I think that it has worked, so far as my experience is concerned, quite well, with a minimum of that kind of interference in regulatory decisions.

This combination of the two responsibilities in one agency has the very much compensating advantage that you have the expertise that is required in your organization in order to discharge the regulatory responsibilities adequately and this is very important in a new and growing technology like this.

That is why we think that the time is not yet right for the separation of the responsibility, and that this advantage far outweighs any potential disadvantage from a possible bias that might be introduced due to the dual responsibility.

REP. HOSMER: I quite agree with your feelings and the facts as the record has displayed them over the years.[19]

Sometimes there is levity:

REP. HOSMER: I think the Turkey Point reactors in Florida are about, what is it, 18 or 22 miles from Miami?

DR. BECK: It is 18 miles. No; I beg your pardon. The low population zone at Florida Power & Light is 5 miles but Miami, I believe, is 18 miles.

REP. HOSMER: I wonder if we could actually consider that a remote location? It depends on which way the wind is blowing, I guess?

DR. BECK: That is quite right.[20]

From all indications, much of the forward propulsion for the nuclear power program seems to have been provided by the joint committee itself. Former Joint Committee Chairman John O. Pastore, senator from Rhode Island, had actually proclaimed himself an "enthusiast" of atomic energy,[21] and other members of the committee have indicated impatience and concern with the rate of progress made in our reactor program. Hence:

REP. HOSMER: Dr. Okrent, you said something that rather disturbs me that you are kind of laying down these academic rules for people coming in for licenses, and you said they ought to be conservative in their design.

I am just wondering how fast we are going to progress if the Committee [the ACRS] is insisting upon this retention of conservatism all the time.[22]

And:

REP. HOLIFIELD: I share the concern of Mr. Hosmer in regard to the length of time required for licensing. It is important because all of these companies are faced with a phenomenal increase in the need for electricity. Naturally, they have to fit it into their schedule of customer need at the end of the 4- or 5-year construction period.[23]

And:

REP. HOSMER: This problem of getting these things closer to big cities is really a tough one. Even if you could get an okay technically from the ACRS, you still have a problem of getting a piece of property somewhere and not making a whole bunch of people mad as they are at Malibu.

I am just wondering every time I go in and out of a big city airport and see all this flat land around there and pieces of property and perhaps the idea might be for the AEC to develop some kind of low profile installation that could take advantage of this already barren land around airports and already mad people, so you would not make anybody else mad, and maybe that might be the area to look.[24]

T. E. Murray, the former AEC member whose remarks we have

quoted in another context, had much to say on the subject of congressional control over the AEC.

> But the Joint Committee's supervisory or "watchdog" powers vis-a-vis the AEC are vague. . . . From time to time, the Joint Committee is called upon to make important decisions, for instance in reporting out the proposed agreements to permit international nuclear cooperation. But when it makes these decisions, it usually has little choice but to make them within a policy framework which for the most part has been constructed by other agencies of the government. . . .
>
> I am not, of course, suggesting that the Congress should be the sole authoritative body to give comprehensive, positive program directives to the AEC, spelling out in precise terms what types of construction projects the Commission should undertake. . . . But it certainly seems to me that within a democratic society the representative legislature must retain the power to make broad policy determinations of a continuing effect and to see to it that they are generally adhered to in the long run.[25]

Thus the internal structure of the Atomic Energy Commission, and the external regulatory function that could be imposed by Congress, have both succumbed to the inexorable thrust of the nuclear power program. The third area where such acquiescence is in evidence is in state jurisdiction.

Although traditionally the responsibility for citizen safety falls within state jurisdiction, as part of the state's general police powers, radiological protection of the citizen has in large measure been preempted by the federal government. Although state and local governments have some say in conventional safety problems in nuclear power plants, and they may regulate some radiation problems under delegation of federal authority in nonpower plant situations, Uncle Sam is the sole boss in power plant radiation safety.[26] The result is that a traditional, and traditionally powerful, defense against the unfair exertion of centralized political power is not operative in the atomic utility field. John Conway, executive director for the JCAE, tried to pin down two AEC members on this point in the 1967 hearings:

> MR. CONWAY: The point has been made, at least the Commission's position has been, that the Federal Government has preempted this area and hence the states do not have the authority to pass laws or impose restrictions relating to radiation protection for nuclear reactors.
>
> MR. PRICE: So far as radiation safety is concerned, that is right.

MR. RAMEY: We don't use the word "preempt" so much. I think we used a different word in the Joint Committee hearings.[27]

Just as so many other key issues came to the fore in the Fermi dispute, the issue of state and local rights was raised at that time as well. Senator Clinton P. Anderson, one of the most active individuals in the campaign to control the atomic energy effort, said at a meeting of the Council of State Governments, May 10, 1957:

> Consider the Dresden reactor. It is to be built 45 miles southwest of Chicago. But is it of no concern to the Governor of Illinois or the Mayor of Chicago? It might explode, it might run away. What happens to the lives of hundreds of thousands, possibly several million, citizens who might be in the direct path of any radiation or fallout from such a catastrophe? Isn't that your problem?
>
> Now suppose that the runaway isn't going to be severe enough to affect the lives of people. There still remains several hundred million dollars worth of property within a 50-mile area of the reactor, and in the direct path of the fallout carried by the prevailing winds.
>
> What do you do to the Merchandise Mart if, through an accident, it becomes dangerously radioactive? . . .
>
> As a comparison of the public interest involved, let us consider this situation: Before the Mayor of Chicago was satisfied that the Prudential Life Insurance Co. would build its new building in Chicago safely and by good standards, he required his representatives to examine the plans before it was built to see if the footings were sufficient for a skyscraper of that size. He can't blindly trust the Prudential Life to build correctly, he must insist on the right of original examination of all the plans.
>
> Likewise, the Governor of Illinois and the Mayor of Chicago ought to insist upon the right of original examination of reactor plans so that they may check them or have them checked by someone familiar with reactor processes to be sure that the reactor will be reasonably safe. Staffs are not yet available for States and cities to do this job completely, but if they continue to abandon this work to the Federal Government on the grounds that the staffs are not yet available, then no staff will ever be available, and as long as the Republic stands, all checking will be in the hands of the Federal Government only. . . .
>
> Big government, given an opening, always takes over. But surely you want to hold on to the things that the State has always looked after, namely, the body—the physical welfare—of the citizens, and their ultimate health and safety.

The issue raised its head again in 1963, after the AEC launched its big push to erect giant reactors around the country, causing Congressman John Saylor to ask the House of Representatives: "Was it actually the intent of Congress to empower the AEC to force a state and/or community to accept a plant, device, or contrivance to which that State and/or community vigorously object? I think not."[28] A number of attempts to establish states' rights in this field, such as a bill introduced by Congressman Leonard Farbstein of New York on January 4, 1965, have been to no avail.

Unfortunately, Congressman Saylor's assertion that many states and communities "vigorously" object to the federal government's attempts to "force" plants upon them is not quite accurate. For although there have been some motions toward acquiring state and local jurisdiction over nuclear plant radiation problems, the states and communities have not always fought "vigorously." Tax inducements, public relations campaigns, and official reassurances have comforted opponents' anxieties. As a result, the issue to this day remains unresolved.

Early in 1965, for example, the New Jersey Board of Public Utility Commissioners ordered public hearings inquiring into the Oyster Creek nuclear plant proposed by Jersey Central Power and Light Company.[29] Later in the same year, the Board asserted some jurisdiction in the matter, noting that while federal law had pre-empted states from imposing regulations with regard to radiation hazards, states retained other protective rights. The Board also recognized that New Jersey state law required it to work for the public interest without regard to distinctions between radiation and nonradiation hazards, and that the constitutional implications of the Atomic Energy Act were not clear and had not yet been resolved.

But the New Jersey Board of Public Utility Commissioners backed off from testing the rights question, explaining it was in "substantial accord" with the AEC on the radiation protection required. Senator Anderson's warning eight years earlier—to the effect that if the states didn't "do" for themselves in the area of radiological protection at atomic power plants, the federal government would—was fulfilled.

In 1971 the issue went all the way to the Supreme Court. Minnesota attempted, in 1968, to enforce radiation release standards 100 times as tight as the AEC's standards. Although the utility did not deny it could meet those stringent standards, it claimed the resultant delays and extra costs were an unfair burden. The Supreme Court ruled against the state, but Minnesota was able to wring a concession from the utility, after the governor recommended a moratorium on new nuclear plants. The utility agreed to cut emissions by 80 percent.[30]

The fourth area in which the Atomic Energy Commission managed to overcome serious opposition is the local community, specifically in the matter of public hearings and public efforts to prevent construction of dangerous reactors in the vicinity of population concentrations. Under Section 189 of the Atomic Energy Act of 1954 as amended, applications for construction permits of nuclear power plants must be given public hearings. We emphasize "as amended" because it was not until 1957, when the secrecy of the AEC's proceedings in the Fermi matter enraged so many congressional members, that public hearings and disclosure of information on safety aspects of reactors was made mandatory.

Since, to all appearances, the "public forum" arrangement is one of the best means, in our democractic system, of counteracting excesses in the behavior of the NRC and proponents of nuclear-powered utilities, one would think that here, at last, the public had the upper hand. Sad to say, that is not the case. The rules under which they are conducted, the issues allowed for discussion, and the persons permitted to intervene or pursue viewpoints opposite from the atomic energy party line are managed by the commission in such a way as to make the term "public hearing" quite appropriate, for it is the public that does most of the hearing and the NRC most of the talking.

Although public enthusiasm, or even approval, is not necessary to the success of a nuclear plant construction application, it is obviously highly desirable. Resentment on the part of the public or its press can initiate a ground swell of apprehension and hostility, which, if not quickly and effectively placated, can turn into a massive, professionally organized resistance leading to a court fight. The further the issues move away from the sanctuary of the controlled public hearing, the greater their exposure before the nation, and the greater the threat to the nuclear power establishment. Although, as we shall see later, hard-sell public relations campaigns undertaken by the utilities are instrumental in softening up public resistance to nuclear reactors, it is on the public hearing that utilities pin their highest hopes of subduing opposition.

Another important reason why it is crucial for opposition to go no further than the hearing is that contested applications play hob with utility construction schedules. When, in congressional hearings,

Representative Craig Hosmer asked Commissioner James Ramey if realistic schedules were being set for the utility companies, he answered, "Yes, but tight," and then added, "And they don't take into account intervention."[31] Tremendous pressure therefore exists to make sure the natives are kept firmly in their place.

The AEC's "Outline for the Conduct of Proceedings by an Atomic Safety and Licensing Board" lays down the ground rules on which the hearing is based, but the rules are unique in that the Atomic Safety and Licensing Board itself, and itself alone, is authorized to decide who shall participate and what the issues will be. "A member of the public does not have the right to participate unless he has been granted the right to intervene as a party or the right of limited appearance for the purpose of making a statement," says the rule book. "The board rules on each request to participate in the hearing on either basis. . . . The important question which may have to be decided is whether a prospective intervenor has such a relation to the subject matter of the proceeding as to warrant granting him leave to intervene." As Harold P. Green pointed out in the *Notre Dame Lawyer* article, "despite the statutory provisions for licensing reactors 'in a goldfish bowl' with public hearings and public disclosure of safety analyses, in actual practice the regulatory procedures tend to stifle public awareness and discussion of the safety issues."

It will be recalled that in 1965 the AEC appointed the Regulatory Review Panel to examine various AEC functions and procedures. This panel, "from outside the Government," recommended that internecine quarrels between AEC safety review groups be suppressed and concealed from the public. It also had some interesting things to say about public hearings, namely, that their basic purposes were: to give the public a "firsthand impression of the applicant's character and competence"; to show the public that the AEC has been "diligent in protecting the public interest" and that the AEC and ACRS staffs "have only the public interest in mind"; to give the public a "convincing demonstration" that a "thorough and competent review" of the applicant's proposal has been undertaken; to develop a factual record in public; and, finally, to provide the public with a "forum for recording its views, both pro and con, on the applicant's proposal."[32] Harold Green summarizes the panel's findings nicely when he says, "Implicit in the Panel's conclusions was the proposition that public hearings are mere window dressing and that the actual decision to license a reactor is made (behind closed doors and beyond public scrutiny) by the ACRS and the AEC's regulatory staff, primarily the latter."

Many proponents of nuclear power, both within the AEC and on the Joint Committee on Atomic Energy, appear to regard the public hearing as a necessary nuisance, held in great measure to give the antinuclear "nuts" their day in court, tolerated largely because it is a good antidote against bad press or the threat of court fights. This attitude can be clearly seen in passages from the 1967 hearings.

For example, in response to a question by Representative Hosmer as to how much mandatory public hearings really accomplished, AEC Commissioner Ramey replied: "I think our experience has been at these hearings that on the first day you do get a sort of representation of the public and of a few individuals and you get other individuals who are sort of hipped on this question of atomic safety attending them."[33] He went on to give an opinion that without mandatory hearings interventions were more likely, and it was agreed by all that once people witnessed the competence of the AEC and utility personnel, they would realize how groundless their fears were. Representative Hosmer summarized the sentiment of the hearing when he said: "I guess we can say that these mandatory hearings are at least a safeguard against these extremist antinuclear people holding a 'kookout.'" According to the record of testimony, this remark was followed by laughter.

The person or group seeking to intervene in the public interest finds the battlefield arrayed with the mightiest resources of the federal government and private atomic industry. The prospective intervenor, unless himself an expert or capable of acquiring the help of experts, is fatally handicapped by his own technical ignorance, and in any event the process of intervention can be prohibitively expensive.

We have seen how, in the case of the attempt by attorney Paul Siegel to intervene in the Turkey Point proceedings on the grounds that the AEC had not considered its responsibilities for reactor sabotage, a prospective intervenor was not even allowed to raise the issues at the public hearing. A similar incident occurred when a group called the Conservation Center attempted to intervene after Consolidated Edison of New York proposed to build "the largest reactor to be considered for licensing to date": Indian Point Unit #2.[34]

The public hearings were held September 14–15, 1966, and there were certainly enough references in the AEC's Division of Reactor

Licensing Safety Evaluation to merit serious public discussion, because the Indian Point #2 *was* the largest reactor ever considered thus far, and it was sited about twenty-four miles from New York City. In addition, the proposed plant had some features that were, in effect, experimental. One was a penetration pressurization system, which, the Safety Evaluation Division stated, "has not been previously proposed for use in other licensed facilities." Another passage declared:

> The American Standards Association and the Institute of Electronic and Electrical Engineers [IEEE] are actively engaged in the development of standards governing the design, testing and installation of reactor protection systems. . . . Evaluation of the Indian Point Unit No. 2 reactor protection system will be based on such standards, as they are proposed or adopted.

The evaluation did not explain what would happen if, after Indian Point #2 was completed, the ASA and the IEEE came up with standards that the reactor failed to meet.

With regard to another safety criterion, the evaluation stated: "Further experimental information should be available from the San Onofre and Connecticut Yankee facilities by the time the Indian Point II facility is to operate." But it did not explain what would happen if those two facilities reported serious inadequacies in that particular safeguard.

Despite these and other chilling deficiencies, the report concluded with what has since become an old, familiar tune: "We believe that these matters can be resolved during the construction of the facility." But Larry Bogart, director of the Conservation Center, did not find this glossing over of the issues any too tranquilizing, and he filed a petition to intervene, listing eighteen objections to the Safety Evaluation.

Unfortunately, Bogart filed a few days late because he had difficulty securing advisers on the safety report. But although the AEC's rule book allows latitude in the filing of petitions to intervene, the commission chose instead to stand on ceremony, and the petition, which would have aired questions of desperate importance to the public, was denied. When Bogart appealed, the AEC told him, "Even if we were to hold that the petition was timely filed, its very general statement of organizational purpose does not set forth an interest of the petitioner in the proceeding which may be affected by Commission action, as required by our Rules."[35] The AEC was referring to the Conservation Center's statement of purpose, which read:

Conservation Center was organized in an effort to help protect the health, welfare and safety of the public in the Hudson River Valley Basin as well as in other areas of the Eastern United States, where blight and pollution are present dangers.

The outcome of the present proceeding, and any increase in levels of radioactivity by the operation of the type of plant proposed, manifestly affects the interest of the petitioner.

Obviously, that purpose wasn't good enough.

Nucleonics could not disguise its delight with this triumph. In its October 1966 issue, it captioned its report "Con Ed Breezes Through Hearing," and noted "one encouraging fact": that utilities, reactor manufacturers, and the AEC regulatory division "have learned to anticipate the potential trouble spots and work together as a team to suppress them through good planning." *Nucleonics* admitted that the Indian Point #2 hearing had "probed only briefly into a few scattered areas for this biggest nuclear plant . . . ever to come up for licensing," then all but declared that that consideration was nowhere near as important as stifling opposition:

> This advance planning [Con Ed and the AEC had of course met on several occasions prior to the public hearing] . . . apparently paid off at the serene hearing. A potential source of trouble evaporated when the petition of the Conservation Center, New York City, represented by Larry Bogart, to intervene in the case in opposition to the project was firmly turned down by the board.

Little has changed in the transmutation from AEC to NRC and from ERDA to DOE. The names are different, but the faces and policies are much the same. The callous disregard for public health and safety continues unabated, even after the near-disasters of Fermi, Browns Ferry, and Three Mile Island. While the application of the National Environmental Policy Act of 1970 has curbed some of the most flagrant abuses, the continuation of America's nuclear program, in the face of overwhelming evidence of the dangers, is convincing proof that the regulators are far too cozy with the regulated. Profit for a few is here gathered at the expense of the rest of us.

An observation made by Frances T. Freeman Jalet about the Fermi hearings holds true of hearings held today, and indeed provides a fitting conclusion to this examination of the Atomic Energy Commission's rise to unlimited power. The commission's rules and regulations, she stated, "obviously were drawn with great care and with meticulous attention to preserve the unfettered freedom of action so uniquely an attribute of the Atomic Energy Commission since its birth. It is as though they constitute a protective shield surrounding the actions of the Commission, making them almost as impervious to attack as are the various containments that shield the reactor itself."

Mrs. Jalet's arrow struck just a little off the mark, however. For if the "various containments that shield the reactor itself" were indeed as impervious to attack as the rules and regulations with which the NRC is clad, this nation would have no cause to be alarmed about the menace of atomic power.

13

The Scandal of Insurance and Subsidies

Go to your strongbox. Open it and take out your homeowner's policy. Run your finger down the column headed PERILS INSURED AGAINST. Do you find anything covering your property for radioactive contamination resulting from accidental discharge of radioactivity from an atomic facility?

No? Perhaps it's elsewhere. Keep going. Ah, *there's* something: "Nuclear Clause." But wait. It says: "This policy does not cover loss or damage caused by nuclear reaction or nuclear radiation or radioactive contamination, all whether directly or indirectly resulting from an insured peril under this policy."

You may want to call your broker to ask whether this clause means that if the reactor going up outside your city has an accident and the fallout contaminates your property, you cannot recover. He will tell you that that's precisely what it means. He may even be aware that Lloyd's of London itself will not write such a policy. He may tell you not to worry, that such an event is highly improbable, but you will wonder why, if it is highly improbable, no one will insure against it. Couldn't an insurer make a fortune insuring against the highly improbable?

Your broker will inform you that although nobody will insure your property against radiation damage, reactor operators are "heavily" insured against claims. A government program will take care of you. Let's take a closer look at the law that created this program, which all homeowners and business interests in the country are to rely on for relief in the event of a nuclear accident.

This law, the Price-Anderson Act, passed in 1957 as an interim answer to the insurance problem and since renewed twice, allows the nuclear power industry to completely bypass the normal insurance and liability process, thus removing a major incentive to keep the plants as safe as possible.

In the 1950s, when the AEC was trying to create interest in a civilian nuclear power generation industry, one major obstacle stood in its way: No one would insure the plants against an accident. When the AEC's first attempt to quantify the consequences of a worst-case accident was released, the situation did not improve for nuclear proponents. That study (the Brookhaven Report, or WASH-740) predicted back in 1957 that $7 billion in property damage could accrue—an amount that would bankrupt even a large insurance company.

The feelings of the insurance industry were well expressed in 1956 by Hubert W. Yount, vice-president of Liberty Mutual Insurance Company. After considering possible consequences of a major nuclear plant accident, Yount was forced to declare to the congressional Joint Committee on Atomic Energy:

> The hazard is new. It differs from anything which our industry has previously been called upon to insure. Its potential is still unknown and must therefore be calculated currently in terms of a body of knowledge which is expanding from day to day. . . .
>
> Very few insurance companies have had any opportunity to develop first-hand knowledge of the problems involved because of the present limited scope of operation. By the same token, very few insurance companies have developed trained technical personnel to assist their underwriting personnel in insurance evaluation of the hazards involved. . . .
>
> The catastrophe hazard is apparently many times as great as anything previously known in industry and therefore poses a major challenge to insurance companies. . . . We have heard estimates of catastrophe potential under the worst possible circumstances running not merely into millions or tens of millions but into hundreds of millions and billions of dollars. *It is a reasonable question of public policy as to whether a hazard of this magnitude should be permitted, if it actually exists.* Obviously there is no principle of insurance that can be applied to a single location where the potential

loss approaches such astronomical proportions. Even if insurance could be found, *there is a serious question whether the amount of damage to persons and property would be worth the possible benefit accruing from atomic development.* [Emphasis added.]

That Yount was not content to confine his remarks to the dubious insurability of atomic power plants, but took it upon himself to question the wisdom of the entire program, is an extraordinary indication of the depth of the insurance industry's reservations.

A limited-liability insurance plan, subsidized to the hilt by the same unfortunate taxpayers who've underwritten the rest of the nuclear industry, was the government's solution. Price-Anderson provides accident coverage of $560 million (roughly one-fourteenth of the Brookhaven estimate) per reactor. There is no recourse for claims above this amount, other than a special allocation of relief funds by Congress—again, out of taxpayers' pockets. When the law was passed, the federal government was responsible for nearly all of the liability —$500 million out of the $560 million in total coverage. The remaining $60 million came from a pool; no *one* company was willing to take such a gamble.

Beginning in the late 1960s, the federal share has declined; private insurers have taken on more of the risk. The dollar amount, though— despite more than twenty years of rampant inflation, and accident estimates running many times higher than the original Brookhaven estimate—remains the same. No claims in excess of $560 million are allowed; plaintiffs will receive but pennies on the dollar. At present, the private share is $490 million, leaving $70 million as the government's responsibility.[2] As it currently stands, this coverage can be broken down as follows: The private insurance pool will provide $140 million. Operators of each operating nuclear power plant (seventy-one, at present) are assessed retroactively, that is, at the time of an accident, rather than paying the premium in advance as with a standard insurance policy—premiums of $5 million. The remainder comes straight from the Federal treasury.[3]

Claims under Price-Anderson are further limited by other factors. According to Senator Mike Gravel (D-Alaska), there is a ten-year statute of limitations; if, in twenty years, a person living in 1979 near the Three Mile Island plant developed cancer, he or she would be unable to sue under Price-Anderson, even if it were possible to prove conclusively that the cancer resulted from the accident.[4] Moreover, to quote Senator Gravel:

> Even with those provisions which supposedly assure $560 million for victims, there are loopholes which protect the industry at the expense of the public. For one thing, industry's costs for investigating and settling claims are to be subtracted from the $560 million. In other words, the funds to pay utility lawyers to challenge victims' claims will come from money supposedly designated for paying those claims. In addition, a utility's property outside of its reactor is covered by the $560 million. This means that the utility will recover from its own "public liability" insurance.[5]

To put it bluntly, utilities, which foisted this unsafe technology upon us in the first place, will be first in line when claims are paid. And with the multibillion-dollar cost of a new nuclear plant, even their prorated share is likely to be the largest slice of a penny-ante pie. Gravel revealed another interesting fact:

> It is also interesting to see how much private insurance the utilities have managed to get for damage to their reactors after an accident. While they claim that only $125 million in coverage can be found for public liability [since increased, as we have seen, to $140 million], their reactors are covered up to $175 million.[6]

An idea of the magnitude of the subsidy provided by Price-Anderson is given by Herbert S. Denenberg, former Pennsylvania insurance commissioner. Working from the updated Brookhaven Report of 1964–65, Denenberg determined that health and property losses in a major accident could total $40.5 billion. With a premium of $580 per million dollars of liability insurance, he calculated that an annual premium for $40.5 billion in liability coverage would equal $23.5 million each year. Current liability premiums under Price-Anderson run well under a half million dollars ($300,000–$400,000), so this subsidy saves each nuclear plant operator about $23 million annually.

Denenberg noted, interestingly, that the annual operating costs of a nuclear plant (excluding various subsidies, but including fuel and maintenance costs) run to $23 million per year.[7] A realistic insurance policy, then, could alone double the cost of operating a nuclear power plant!

From the very beginning, the Price-Anderson plan has come under fire. Even some of the AEC staffers thought it was unwise to use this plan, primarily because of concern over its constitutionality.[8] Congressman Robert Secrest, speaking on the House floor ten years later, again raised the issue of constitutionality, before the Act was first renewed:

> It hardly seems proper to assume that the bankruptcy power gives Congress the authority to completely eliminate financial responsibility for damages without even touching the assets of the utility involved. . . .
>   It does appear that Congress has jurisdiction over the construction and operation of atomic power plants because of the commerce clause. Even so, the delegation of jurisdiction under the clause does not give us complete constitutional authority to act as we wish, without limitation. Whatever laws we pass under the commerce clause will not be constitutional unless they are reasonable and appropriate means to a lawful end. There is very serious question as to whether or not an act of Congress giving complete immunity from responsibility for negligency constitutes a reasonable and appropriate means to a lawful end.[9]

At the same time, utility officials, as they have done all along, admitted that the nuclear industry could not exist without Price-Anderson. Philip Sporn, an expert on nuclear economics states: "Price-Anderson insurance is absolutely essential to the continued development of nuclear power. Its elimination would choke off the construction of nuclear power plants."[10] If nuclear power is not insurable, it is reasonable to assume that it isn't safe. Otherwise, insurers would jump in to reap large premiums protecting against "impossible" accidents.

Price-Anderson has been challenged both on the floor of Congress — though the Act has been renewed through 1987 — and in court. Judge James B. McMillan, sitting in North Carolina Federal District Court, found the Act unconstitutional on March 31, 1977 (almost exactly two years before the Three Mile Island accident), claiming it was in violation of the Fifth Amendment. Although the utility, Duke Power Company — and thus the Act's constitutionality — was upheld on appeal to the Supreme Court, which decided in favor of the utility in June 1978, this is sure to be raised again.[11]

More recently, in the wake of Three Mile Island, numerous suits have been introduced, the combined dollar value of which far exceeds the arbitrary $560-million limit. In the first two weeks after the accident, Metropolitan Edison paid out expenses of 2,500 families who evacuated the area, running to $900,000. On top of this, several class-action suits have been introduced,[12] one of which seeks damages for each person who lives or has a business within twenty-five miles of the complex and asks that the utility be responsible for providing twenty years of medical coverage for the affected parties.[13]

Meanwhile, another congressional attempt is being made to bypass Price-Anderson. Congressman Ted Weiss (D-NY) has introduced a bill to make utility companies liable for nuclear accidents up to the total amount of their assets.[14] Congressman Morris Udall (D-Ariz), who ran for President in 1976 on a strongly environmentalist platform, is chairing the committee that will examine Weiss's bill as part of a general inquiry into nuclear safety sparked by the Pennsylvania accident.[15]

Citizen referenda have also addressed the Price-Anderson question. Among the provisions on a Montana bill passed by referendum in 1978 is one providing for full liability in case of an accident. This and other provisions will essentially ban nuclear plant construction in that state. The nuclear industry may be forced to justify its claims of safety by accepting the risks if their executives are wrong.

Opponents have strangely neglected one important argument against the nuclear industry: the devastating risk it entails for *every other industry in the country*. A single major accident (with release of sizable quantities of radioactivity) at a nuclear power plant anywhere in the United States would mean almost certain financial ruin for other business interests in the vicinity. As Senator Gravel expressed this in listing business and civic groups as potential allies for nuclear opponents:

> When those in business understand that a single nuclear accident can do twice the property damage of Hurricane Agnes, totally ruin the economy of a huge region of this country, and lead to an unplanned nuclear shutdown from coast to coast, they may consider an immediate moratorium the *only* responsible position. . . .
>
> The country's farmers all stand to lose a great deal from nuclear electricity. The AEC's own "Brookhaven Report" estimates that one major nuclear accident (or act of sabotage) could radioactively contaminate 150,000 square miles of agricultural land, which is the equivalent of a square with one side reaching from Chicago to Cincinnati or from New York City to Richmond, Virginia.[16]

We have already quoted Hubert Yount, vice-president of the Liberty Mutual Insurance Company in 1956, on the over-all philosophy of government insurance for nuclear plants. However, Yount made an additional point:

> We believe that your committee should have clearly in mind the nature of the third party losses likely to arise. The first category of such losses are those arising from damage to persons and property of individuals and corporations under which claims will be filed directly by the injured persons involved. A second category of losses will arise as subrogation claims filed by insurance companies who have been compelled to pay losses under their policies and in turn pursue their rights at law to collect from the primary interest responsible for the loss.
>
> A good illustration of such a subrogation loss would arise when radiation produces injury, damage or death to the employees of another corporation which losses are paid under workmen's compensation policies. These injuries would then become the basis for third-party claims, based upon negligence, against the operator of the reactor and other interests involved.[17]

Later in the same testimony Yount asked:

> Would the establishment of maximum legal liability through legislation, if valid, remove the pressure to conduct a safe operation? Or at least furnish an incentive to a less safe operation? And how can you deal in this manner with the numbers of insurable interests involved in connection with each installation?
>
> We have no answers to these questions. They are listed to indicate some of the reasons why we object to such proposals.[18]

Twenty-three years later Jack Anderson, writing in his column of May 11, 1979, stated:

> If the Harrisburg area, or any other community near a nuclear power plant, had been exposed to a fullblown nuclear disaster, it would have meant financial ruin for thousands of residents.
>
> The reason: Federal law sets a limit on insurance compensation for nuclear accidents, and the limit is ridiculously low. . . .
>
> Here's how this inequitable insurance compensation (Price-Anderson Act) might have worked out in the Three Mile Island situation, according to an internal study prepared by the Federal Insurance Administration:
>
> Assuming that serious radiation is confined to a 20-mile radius, 200,000 dwellings worth an average of $50,000. Each would be rendered permanently unfit to live in. Adding in an estimated $5,000 in relocation expenses and $12,500 in personal property loss for each family, plus damage to area businesses, the total loss would amount to $16.8 billion.

The study also assumed that people would be evacuated and no medical expenses or damages would be incurred.

The average family, losing a total of $67,500, would get just $2,247 out of the $560 million kitty. This would amount to just over 3 cents of compensation for every dollar loss.[19]

Negative effects on property values have already occurred. Another Anderson column, in November 1977, entitled "Cloud Over Pennsylvania," stated:

> The Shippingport people are frightened and alarmed. Residents have filed a lawsuit against CAPCO. They charge that the power plants have disrupted their lives, ruined their property and destroyed the peaceful aura of their beloved Beaver Valley. *Property values have plummeted 40 percent.* No one will touch the land near the nuclear site.[20]

This potential for drastic reduction—or even total elimination—of real-estate values in the vicinity of nuclear power plants should offset the enthusiasm for the property-tax advantages such plants often bring to their host communities. In fact, sheer lack of awareness seems to be the only logical explanation for the failure of the business community and individual homeowners to rise up against such unprecedented threats to their very survival. No other explanation seems plausible for the large (and politically powerful) industry leaders who seem willing to share with the public the pitifully inadequate government insurance now available.

To add insult to injury in early August 1979, according to a report by Joanne Omang in the *Washington Post,* "The Nuclear Regulatory Commission (NRC) has tentatively decided that the Three Mile Island nuclear power plant accident was *not an extraordinary event.*

"If the finding stands, it will make it much harder for people who claim millions of dollars damages from the March 28 incident to press their cases in court."[21] It is unprecedented folly and arrogance to permit one industry to endanger every other industry in the country.

The Atomic Energy Act of 1946 stipulates that whenever, in the opinion of the Atomic Energy Commission, nuclear power has been sufficiently developed to be "of practical value," the commission is obliged to report to that effect to the President so that the political, economic, and social implications can be evaluated and acted upon.

Although the AEC had often been requested to do this for the commercial nuclear program, it wasn't until 1971 that they stopped classifying all reactors as experiments. Until that time, utilities were benefiting from several large-scale government subsidies in research and development, in fuel buy-back guarantees and in other areas, *in addition* to the huge subsidies that still exist.

In 1976 nuclear plants generated (gross output) 192-billion kilowatt hours, while federal research monies for nonmilitary plants totaled $1.3 billion. This amounts to a hidden additional cost of 7 mills per kilowatt hour for that year. And until 1974 nuclear sources were the only energy generating methods subsidized by federal research money; now, other sources combined add up to a miniscule portion of the budget.[22]

Research and development, however, is only one of the federal subsidies. Government uranium enrichment plants charged, until 1977, only about one-third of what a commercial plant would have charged. These subsidies at the front of the nuclear fuel cycle add about 1.2 cents per kilowatt—to the stated cost of 4 cents. In other words, these three subsidies served to cover up a more than 25-percent increase in the actual cost of nuclear power.[23] When one considers that nuclear and coal plants are in the same economic ballpark, this becomes extremely significant. Even disregarding subsidies, the cost of nuclear can run as high as 36 percent more than the cost of coal, exclusive of the tax bill that will come later.[24] To further complicate the nuclear picture, a sharp look must be taken at the economics of the end of the nuclear fuel cycle—reprocessing, waste storage and eventual disposal, and plant decommissioning. All of this will cost a pretty penny, but since standard procedure has been to deal with these problems only when forced to, the true costs remain a mystery.

Nuclear fuel becomes highly contaminated and must be replaced long before it is fully consumed. It is thus possible to retrieve fissionable substances from the spent fuel. It is also possible to make nuclear bombs in this manner. Reprocessing has so far been an economic disaster, with the government picking up most of the tab. Because of the security consideration, President Carter has moved to limit reprocessing, both here and overseas. Its current future is in doubt. Wastes, meanwhile, continue to be stored in temporary holding tanks, many of which are now critically overcrowded. If much more waste is added to some of these pools, the possibility of reaching critical mass leading to an uncontrolled chain reaction becomes far too likely.

We will not be the ones to pay for cleaning up all of the waste mess. Rather, our children and grandchildren on to infinity will be saddled with the job—assuming the rather doubtful prospect that a solution of any permanence is ever found. This is the very real cost of nuclear power, and it will of necessity amount to many trillions of dollars over the centuries. Temporary storage of nuclear waste is not cheap either, as we have seen. And another unconsidered cost is that of decommissioning.

A nuclear plant has an expected lifetime of thirty to forty years, although in practice many plants have been closed at far younger ages. Vallecitos in California was abandoned after six years;[25] Indian Point #1 was closed because it lacked essential safety devices after a comparable period of time; Hallam, Nebraska,[26] Enrico Fermi #1, and Three Mile Island #2 all operated for only a few months before being shut down.

All of these plants then become part of the radioactive waste problem. To safely decommission these plants will take large amounts of capital; at least one plant cost as much to decommission as to build.[27] The problems are complicated because one cannot easily move a 1,000-megawatt plant as if it were a fuel bundle. The problem is even more severe in a plant such as Fermi or Three Mile Island, where a serious accident has caused the plant to become intensely radioactive. The taxpayers will have to cover these costs, because utilities have not set aside any money for this. The first attempt to address this question was made in 1978, when Montana passed a referendum that, among its other clauses, requires capital to be set aside for this purpose.

So much for the government subsidies. This, however, is by no means the end of the nuclear economics story. It should be emphasized again that the nuclear process is extremely capital-intensive, creating fewer jobs per dollar than almost any segment of the economy. Furthermore, the gross energy of nuclear plants is used to power sophisticated machines that replace workers; nuclear power is geared, in short, not to the residential user, but to the automated factory.

The cost of actually constructing a nuclear plant is nothing to sneeze at, and virtually every plant in existence or under construction has been saddled with enormous cost overruns. In other words, when a utility announces a nuclear plant, its projected cost is misleadingly low. Overruns were bad enough in the early plants: Hallam, Nebraska, was originally projected at $67 million; the true cost (of which the government paid three-quarters) was $84 million. Indian Point #1 in New York cost $134 million instead of $55 million, nearly two-and-one-half times the original estimate. The cost of Fermi #1 soared from $62 million to $109 million.[28] And as plants became larger, so did the overruns. A 149-percent overrun raised the cost of Georgia Power's Hatch plant from $111 million to $276 million; the Browns Ferry complex in Alabama went from $247 million to $513 million, an overrun of 168 percent.[29]

In part because of retrofitting safety devices onto an obsolete reactor design, cost overruns are jacking up the price of the Shoreham plant, now under construction on Long Island's North Shore, almost beyond belief. The 820-megawatt plant was estimated in the late 1960s at $271 million. Now nearing completion, the plant has cost $1.2 *billion* ($1,200 million)—*almost five times as much* as the original estimate. Small wonder, then, that Long Islanders' bills have almost tripled from 1970 to 1978. Long Island Lighting Company (LILCO), meanwhile, has again asked for another rate increase to pay for Shoreham, this time $171 million.[30]

Another important component in nuclear economics is capacity factors, the percentage of potential output that a plant actually generates. Capacity factors can thus be expressed as fractions—actual production over maximum total capacity. The AEC used in 1974 a figure of 75 to 80 percent. David D. Comey looked closely at this data and came up with a figure of 57.3 percent for 1973.[31] The newer, larger plants—that is, those larger than 1,000 megawatts—showed capacity factors of 44.5 percent. This would mean that *two 1,000-megawatt plants would be required to produce 890 megawatts of actual power.* By comparison, the capacity factor for all coal plants between 1964 and 1974 was a far better 68.9 percent.[32]

Since in the case of nuclear the energy output must be balanced against the energy needed along all the various phases of the fuel cycle, this is extremely significant. Coal, once mined and cleaned, can simply be crushed, taken to a power plant and burned. As we have seen, once uranium is mined, it must be milled and enriched. *Enrichment takes vast amounts of energy; the three plants in the United States (in Ohio, Kentucky, and Tennessee) consume fully 3 percent of all electricity produced in this country*—to say nothing of the millions of gallons of water per day that are sacrificed to cool these plants.[33] Once enriched, the uranium is fabricated into fuel rods, consuming more money and energy, before it finally arrives at the nuclear plant. Once fissioned, the fuel must go through all the processes at the back end of the fuel cycle, unlike coal.

When nuclear plants operate too far below their expected rate of performance, they become billion-dollar white elephants, consuming more energy than they create and chewing up vast sums of capital. And of course as top-quality uranium reserves are drained off, it will take more money and energy to mine low-quality uranium, which further increases the cost to the consumer. Nuclear power, in other words, has been the recipient of an energy subsidy—as well as a financial subsidy—of truly massive proportions.

14

From Fissile to Fizzle

In 1969, when uranium could be obtained for only eight dollars per pound,[1] it was predicted that a uranium shortage would begin to be felt around 1980. This prediction, it turns out, erred greatly on the side of caution; there has been a serious shortage of uranium for several years now. The current price for a pound of unenriched uranium stands at $43.00[2] — an increase of almost 550 percent. Even compensating for inflation, it is obvious that the price of uranium is headed for the skies.

To make the purchase of its reactors more attractive, Westinghouse had guaranteed to provide twenty-seven utilities with 40,500 *tons* of uranium at $8–10/lb., with adjustments for inflation; the company's own holdings amounted to only 7,500 tons. To fill the contracts, Westinghouse would have had to pay $2.6 billion for uranium that would be sold for only $594 million — a $2-billion difference.[3]

In 1975 Westinghouse broke the contracts; it was promptly sued by its customers.[4] Acting on information leaked by Australian environmentalists and released by Friends of the Earth, Westinghouse responded with a suit of its own, charging Gulf Oil and twenty-eight other companies with collusion to inflate the price of the fuel, and further charging that the companies had boycotted Westinghouse, re-fusing to allow it to buy more uranium.[5] Although the price of uranium has now leveled off — it had already reached $41/lb. by 1977 — it can be expected to begin rising once more. Bids in 1977 for uranium available in 1980 had been placed at $52/lb.[6]

While the shrinkage in reactor orders (from a high of thirty-five in 1973 to a bare handful in the last few years)[7] had lessened the de-mand, the inescapable fact is that we are running out of uranium. In E. F. Schumacher's terms, we have been squandering nonrenewable resources (including both uranium and fossil fuels) as if they would last forever; we have used this "natural capital" as if it were income.

Considering this, it is not surprising that uranium interests have failed to learn from their experiences with oil, gas, and coal. Of the twenty largest corporations holding domestic uranium reserves, fully eight — including three of the top four — are major oil companies, such as Gulf and Exxon; most of the others have other large energy and re-source mining interests.[8] Oil companies, in addition to their 47-per-cent control over domestic uranium,[9] have acquired major uranium interests overseas.

As with virtually every other resource developed under the pres-ent economic system, the first to grab gets the biggest, most acces-sible, and most profitable chunk of the pie. Uranium lodes containing the highest percentage of fissionable uranium 235 — which exists in very small quantities in any case — and located closest to the surface have of course been the first ones mined. The good stuff is already gone, and even without the operation of an OPEC-style cartel, it will cost significantly more to produce the same amount of energy. In-creased costs of mining, further problems with finding places to put the tailings (the radioactive ore from which uranium has been ex-tracted), additional time and energy spent in enriching the concentra-tion of uranium 235 to a fissionable 3 percent — a process that already requires 1,700 passes through the enrichment equipment[10] — can only add to the costs.

"A ten dollar per pound increase in uranium oxide leads to a tenth of a cent (one mill) per kilowatt-hour increase in the price of de-livered nuclear electricity and even the Atomic Industrial Forum, a nuclear industry association, claims that nuclear power has only a 3.2 mill per kilowatt hour advantage over coal."[11]

As we have seen, critics claim that nuclear power, subsidies and

all, is still 36 percent more expensive than coal. A look at uranium reserves and resources, however, proves the folly of ignoring this, which
nuclear proponents have done.

Known uranium reserves (as of 1977) are 690,000 tons, including
some 90,000 tons produced as byproducts of phosphate and copper
mining. Potential reserves—resources of known size, whose exploita-
tion will become cost-competitive as uranium increases in price, as
well as hypothesized, undiscovered deposits—total 3,760,000 tons.
Most of this is uranium of an extremely low grade, to the extent that
its extraction would cost at least one hundred dollars per pound. This
is only the cost, not the *price;* allowing the miners a profit would
further boost the price.

Realistically, then, we must concern ourselves only with the reserves—that 690,000 tons of uranium ore waiting to be mined. The
combination of high price and low yield puts the low-grade ore out of
reach essentially forever. For example, in ore containing 70 parts per
million uranium, one ton would contain 2.24 *ounces* of uranium. Of
this already scant figure, another 30 percent is lost to tailings. Seven-
tenths of one percent of this will be fissionable uranium 235, which is
the standard ratio of uranium 235 to the vastly more plentiful uranium
238. About another one-third is lost in enrichment, and nearly 15 percent of the pittance remaining will fuse into uranium 236. Furthermore, only 70 percent of this bit of dust will be burned as fuel before
the reactor accumulates too many fission products, forcing replacement of the fuel. Therefore, to quote John J. Berger in *Nuclear Power:
The Unviable Option:*

> Out of the original ton of rock mined, only a grand total of *forty-four ten
> thousandths (0.0044) of an ounce* of uranium-235 can actually fission to pro-
> duce electricity, supplemented by an "energy bonus" of about 43 percent
> from the fissioning of other isotopes. . . . The net energy so derived from
> such low-grade uranium is . . . 836 kwh.[12] [Emphasis added.]

Berger goes on to suggest that such uranium would cost $172 per
pound of oxide. This of course is four times the current price, and
seventeen to twenty times the fuel prices in Westinghouse's original
agreements. Using data worked out by Morgan G. Huntington, an engineer for the Department of the Interior, Berger claims the 690,000
tons in reserve are sufficient to fuel only sixty-two 1,000-megawatt
reactors over their forty-year operating periods. A careful look at the

figures, however, makes it obvious that even Berger is being generous. Using Huntington's figure of 22-million kilowatt-hours gross yield per ton of uranium, he points out that this was the maximum performance. Actual performance has been more like 6-million kilowatt-hours per ton.

A simple mathematical extrapolation of this data yields an astounding result. If Berger's and Huntington's figures are correct, the true figure is less than what's already on line: *Seventeen 1,000-megawatt reactors could be fueled for forty years* with 690,000 tons of domestic uranium reserves. Period.

By contrast, the AEC of the Nixon years was predicting one thousand operating reactors by the turn of the century. Where the fuel would come from is a matter of pure speculation. This of course is based only on the gross, and this gross, according to Berger's optimistic figures, could supply at most 5 percent of U.S. energy needs.

While 10.5 quads of energy have already been consumed by just the enrichment phase, this amount of uranium cannot be expected to produce a gross of more than 3.91 quads. In yet another way is atomic energy a losing proposition.[13]

Nor can we rely on foreign sources. Cries for energy independence are going to look very foolish when uttered by nuclear proponents, who would switch us from dependence on Arab and Latin American petroleum to South African, Australian, and Canadian uranium. Moreover, our great efforts to export reactor technology have created a foreign market for uranium that exceeds demand—a market that would have been of interest only to the United States, had it not been so insistent on spreading the poisonous wealth around.

The only other possibility is the breeder, and that is a dismal prospect indeed. For breeders to be most effective, as producers both of electricity and of fuel, they must each have a capacity in the 1,000-megawatt range. Enrico Fermi, the only commercial breeder to operate in this country, was operating at only 11,000 *kilowatts* (a megawatt is a thousand kilowatts) at the time of its near-calamitous accident.[14] The risks, in short, are great.

And, as previously noted, the French breeder program has been a dud, producing far less plutonium than had been hoped (or feared, depending on one's point of view). Even with an all-out effort, the chances of a successful breeder program in operation by the end of the century are essentially nil. The current fuel-wasting generation of reactors, in other words, may well be both the beginning and the end of nuclear power in our society. When these are gone, the next generation will have to look elsewhere for fuel.

15

Energy Alternatives

It is difficult to choose which of the many myths regarding nuclear power is most frustrating to the citizen who has discovered the truth about this uniquely hazardous and expensive source of energy. For those who value the gift of life and to whom the health of this and future generations is of greatest importance, while the myths regarding safety must be of primary concern, the myth that nuclear power plants are *necessary* — the myth that there are no viable alternatives available at present sufficient to meet the nation's needs and to maintain its industry and standard of living — must run a close second.

That this is a myth — one that is often perpetuated by nuclear proponents (sometimes out of ignorance, more often because of self-interest) — can be documented by numerous government and industry reports and by several excellent books on various alternatives to nuclear power including *Rays of Hope: The Transition to a Post-Petroleum World* by Denis Hayes (W. W. Norton, 1977). In addition, many anti-nuclear books, such as *No Nukes* by Anna Gyorgy and friends, have excellent sections devoted to the vital subject of infinitely safer, more reliable and less expensive energy alternatives. The following brief rundown of some of these alternatives is intended to indicate how extensive and varied our resources are—and to give some idea of the numerous approaches which have already been proposed to implement them.

A report released in April 1979 by the Office of Technology Assessment said that the U.S. could save 19 to 29 billion barrels of oil by the year 2000 through *conservation*. This is the figure given if *homeowners* got maximum energy (and dollar savings) by full use of *available* conservation techniques. In an age when nuclear proponents so frequently threaten the public with an image of "freezing in the dark" (if we forego nuclear power) it is interesting to note that this report emphasized that no adverse lifestyle change would be required. The use of better insulation, storm windows, and weather stripping, plus more efficient heating and cooling devices would insure quite comfortable living—but at maximum savings of both energy *and* money. Indeed, since it has been estimated that 40 percent of all low-income families lack insulation in their homes, the poorest American families are currently paying much more than necessary for heat —or doing without.

But if effective and readily available conservation methods can be an economic boon for the average homeowner, it can be equally important to all of us as taxpayers and consumers. The Senate Commerce Committee estimated in late 1976, for example, that an energy-efficiency program upgrading public schools, hospitals, government office buildings, and so on could save American consumers many millions of dollars a year—and generate more than 100,000 new jobs.

Industry, which consumes a huge 40 percent of the nation's energy, has already begun to discover and profit from available conservation methods. Large companies, such as Eastman Kodak's headquarters and IBM have been reported to have cut energy use by 38 and 35 percent respectively. An ad for IBM reported that one of its computers, which automatically shuts down part of the *Philadelphia Inquirer* and *Daily News* plants' ventilation systems during start-up of the huge presses, has cut power consumption by 20.7 percent in its first twelve months and paid for itself within a year and a half in power saved.

According to DuPont's G. Frank Moore, a "significant conservation effort" at any plant could reduce consumption by 7 to 15 percent. And according to his estimates, a 10 percent industry-wide cut in energy use, would save the equivalent of 1.5 million barrels of oil a day.

Even greater industrial conservation is possible as industry learns to use its heat, trash, waste steam, and gas to generate its own electricity. According to Charles A. Berg, former chief at the Federal Power Commission, "An average paper plant could produce three to four times as much electrical power as it could consume."

What to do with *excess* electricity would then become a problem —but one that could be solved by modifying regulations that now prevent anyone other than utility companies from buying or selling bulk electrical power. Persuading utilities to buy surplus industrial electrical power would give the paper, glass, and other big energy wasting industries an incentive to make use of the power now being wasted.

Conservation is an alternative with many applications. According to an American Institute of Architects report, "Energy and the Built Environment," currently available conservation techniques for existing and new building construction could provide a savings by 1990 of 12.5 million barrels of oil a day. A new light bulb being developed by General Electric would use only one-third of the energy currently consumed by bulbs. In October, 1979, Borg-Warner Corporation unveiled a new heat pump with an inverter that has been documented to save 17 percent in electricity. The system, termed by the company "the most energy-efficient home heating and cooling system yet developed," is slated to be on the market in early 1980.

These three examples demonstrate that we have more than an adequate number of ways to conserve large amounts of electrical energy (and gain a degree of independence from foreign oil and total independence from nuclear power).

Apart from energy now being wasted, there is another resource in abundance in the U.S.—*garbage.* The advantages of using garbage or refuse as a fuel source are many—not the least of which is the enormous cost of present methods of garbage disposal. Current garbage disposal methods also present growing health hazards due to the already inadequate landfill space. And in large cities there is the air pollution problem posed by thousands of individual incinerators in large apartment buildings.

An article in the *New York Times* in June 1977 pointed out:

> Over 120 trash-to-energy plants have been built around the world, primarily in Europe, to help meet urban energy demands. Only a few exist in this country. One has been operating in Saugus, Mass., for over a year. It proc-

esses 1,600 tons of trash daily from twelve communities, plus two districts in Boston, and sells the steam it produces to a nearby factory, resulting in a "savings" of 73,000 gallons of fuel oil daily. Ten American companies are ready to build such systems, which could save cities millions of dollars in garbage disposal costs and supply 10 to 20 percent of their electrical needs.

Another underused resource was described by David E. Lilienthal in an article for the *Smithsonian* in September 1977. Mr. Lilienthal advocated putting our rivers to work to supply electricity. In his article he referred to a Federal Power Commission study which found that "if only ten percent of the hydro-power potential of our 49,000 dams were even partially developed, we could save the energy equivalent of 180 million barrels of oil every year."

It is significant in view of the growing dependence on unreliable and expensive nuclear plants in the New England states that the Federal Energy Regulatory Commission has cited 228 small hydro-plants that have been *abandoned* in New England.

With regard to the country as a whole Mr. Lilienthal stated:

> The FERC estimates that only about 1400 of the nation's 49,000 dams have been developed for power generation; the rest were built for river navigation, flood control, irrigation and recreation. There are 343 flood-control dams in the Northeast, (229 in New England alone); many can be made to accommodate power installation without compromising their original function. . . .
>
> In communities around the country, many of the small- and medium-sized projects can be developed at lower capital costs per unit, and will produce energy at lower production costs per unit, than we are likely to get from huge generating stations using less permanent, less reliable, more hazardous resources. Moreover, they can be built quickly, compared to the 10 or 12 years required nowadays to design, license and build a large coal-burning or nuclear plant.

The good news about alternatives grows more impressive. every day. In early May 1979 an Associated Press report stated, "The U.S. Geological Survey reported Wednesday that the United States could tap as much energy from underground heat sources as current oil consumption would supply in 162 years."

In an article on wind turbines in August 1979, the *Chicago Tribune* made encouraging observations.

> Darrell H. Baldwin, NASA-Lewis expert, sees the first working wind farms in the United States being built about 1985.

"Eventually, say by the year 2000, the nation could reasonably expect to be producing 5 to 10 percent of its electricity by wind turbines," Baldwin said. . . .

Loew [director of the Boeing wind energy program] said the rapidly advancing research will bring the kilowatt-hour cost down to about 4 cents, which would compete with commercial generation in "certain areas" of the country. The cost of power in New England has been estimated at about 6 cents per kilowatt hour. . . .

Many parts of the country have enough wind to make wind turbines practical, NASA and DOE engineers say. Top areas include New England, the two coasts, and the Plains.

One of the most knowledgeable people in the country about alternative energy sources and particularly solar energy is Denis Hayes, author of *Rays of Hope*. In an article based on his book for *Critical Mass Journal's Nuclear Power Primer*, Mr. Hayes wrote:

The most exciting solar electric prospect is the photovoltaic cell—now the principal power source of space satellites and the main element in photographic light meters. Such cells generate electricity directly when sunlight falls on them. They have no moving parts, consume no fuel, produce no pollution, operate at environmental temperatures, have long lifetimes, require little maintenance and can be fashioned from silicon, the second most abundant element in the earth's crust.

Photovoltaic cells are modular by nature, and little is to be gained by grouping large masses of cells at a single collection site. On the contrary, the technology is most sensibly applied in a decentralized fashion—perhaps incorporated in the roofs of buildings—to minimize transmission and storage problems. With decentralized use, solar cells can be combined with compatible technologies to use waste heat for space heating and cooling, water heating and refrigeration.

The manufacture of photovoltaic cells is currently a low-volume business—only 750 kilowatts of photovoltaic capacity were produced in 1977—and the products are consequently rather expensive. But a recent U.N. report concluded that solar cells would become cheaper than nuclear power if they received a total investment of $1 billion—less than the cost of just one large nuclear power plant.

Writing about another aspect of solar power, Mr. Hayes noted that heating water with sunlight is simple. "Other countries are outpacing the United States in this field. About 30,000 American homes heat their water with sunlight. In tiny Israel, 200,000 homes have solar water heaters, and in Japan the figure is over 2 million."

For those who are still concerned that a drastic shift to solar energy might have a negative effect on jobs and the economy, a UPI report of July 29, 1979 included the following: "A recent congressional study reached the startling conclusion that a race-to-the moon type solar program to replace traditional energy sources would make up for lost employment in the coal and oil industry with 3 million jobs to spare."

In our effort to promote alternative sources of energy we cannot ignore economic realities. There is a practical need, (even a moral obligation), to offer some form of indemnification to corporations which have huge amounts of money invested in nuclear technology.

Senator Gravel has recognized this and tried—unsuccessfully—to provide such indemnification. A press release from the Senator's office, for example, asserted:

> This country could declare a moratorium tomorrow—and we might even hear a sigh of relief from some worried people inside the nuclear establishment—provided we insured jobs and offered indemnification.
> After all, we pay landowners billions of dollars every year not to grow crops; we can certainly afford to pay people not to build radioactive machines which could contaminate an area reaching from New York City to Richmond, Virginia.
> If we take care of the financial hardships of a moratorium, arguments in favor of nuclear electricity may lose some of their frenzy.

We are suggesting a major shifting of national priorities away from an area of substantial investment. But one point should not be overlooked: this investment includes the arbitrarily high value of uranium and plutonium, a value placed because of the *assumption* that these would be essential materials for the energy industry. In addition, some of the actual investment—in reactors and the buildings housing them—could be converted. Hence these do not have to be viewed as total losses to either the nuclear industry or our government.

An even more vital economic fact is this: a moratorium on future nuclear power plants, combined with a phasing out of those which already exist, *means huge savings in the future from several sources which will not be needed if nuclear plants are discontinued.*

Again, Senator Gravel cited:

> Savings from an indefinite moratorium on atomic power plants would also include the following:
> The extra cost of having to dismantle 800 radioactive powerplants at the end of their 40-year service;

The budgets—40 years at least—which would be required to build and operate Federal, State, and local monitoring networks for radioactive pollutants;

The budgets . . . for . . . licensing and regulatory and litigation staffs;

The budgets . . . for . . . research program on radioactive pollutants in the environment;

The huge "contributions" from taxpayers for nuclear breeder research and demonstration plants;

The huge costs of improving and expanding uranium enrichment facilities. . . .

Senator Gravel went on to name a few more. Also, we must include the enormous and continual expense of safeguarding all the *additional* quantities of radioactive wastes which a continuing nuclear program would entail.

It is clear then that taxpayers have the choice of continuing to subsidize an industry which jeopardizes lives, health, and property—and every other industry—or supporting an indemnification program which would include conversion to alternative technologies.

16
Citizen Action

People who are convinced that nuclear power plants represent unique, unprecedented, and unnecessary hazards to their lives, health, children, and property have many facts and experiences to support their view. Unfortunately, the merit of their arguments is not necessarily enough to win the fight against a powerful nuclear industry. It is, therefore, essential for citizens to work toward an effective grassroots movement.

There are many antinuclear groups throughout the country — and the world. Methods of the various groups differ. Some concentrate on educational projects; others focus on legislative action, public demonstrations (including the picketing of utilities), or, in the case of the Clamshell Alliance, the "occupation" of a proposed nuclear power plant site.

Taking Issue with the Economics

One of the first lessons any realistic opponent learns is that the name of the game is money. Insistence that the nuclear and utility industries put their own money on the line, pledging this as "security" for the plants they insist are safe, is a first—and vital—step toward making the nuclear power industry more responsible. Whether it is to be a permanent commitment or a temporary one should depend on the willingness of such corporations to work together with the government on an indemnification program acceptable to all concerned. But one way or the other, the endangered private citizen and taxpayer should no longer tolerate being forced to carry the financial burden of insurance against nuclear catastrophe.

To make sure that everyone concerned gets the message, there are immediate steps citizens can take as individuals and as part of a grassroots movement. Nuclear opponents can and should use their *own* money—as customers of both reactor manufacturers and utility companies—in a way calculated to take the profit out of nuclear power. A boycott on all other products of reactor manufacturers such as Westinghouse and General Electric is one impressive way of saying, "We don't want any more nukes!" Utility customers can enclose anti-nuclear messages with bill payments and can write to the companies' presidents and board chairmen expressing a willingness to cut electricity use to the bare bones, until nuclear plants are shut down once and for all. If large numbers of consumers do this—and carry out their resolve—electric companies will feel the pressure where it hurts. More novel actions have been proposed by some groups—one day a week electricity boycotts (with candlelit windows signifying participation) and mass campaigns to withhold payment of a portion of the electricity bill.

But if this is not enough to persuade private utilities, a campaign for municipal public power (accountable to voters, not stockholders) to replace such utilities would be in order. (Many additional ideas and much useful information can be found in two Environmental Action Foundation booklets, "How to Challenge Your Local Electric Utility" and "Taking Charge: A New Look at Public Power.")

Consumer pressure and stockholder action can let big and small businesses of all kinds know that uninsured losses from nuclear power accidents can mean bankruptcy for them. A particularly pertinent argument which can be emphasized is that under the Price-Anderson Act, businesses can only recover a small percentage of claims against losses in the event of a major nuclear accident. The risk of such an accident is substantially higher than businesses have been led to believe.

Stockholders in utilities, in reactor manufacturing companies, or in any industry can take individual action. A *New York Times* report on the 1970 formation of the Council on Economic Priorities quoted Alice Tepper on the effectiveness of such action.

> Investors, students and consumers concerned with social issues are taking a new interest in the policies and practices of corporations. Consumer boycotts, selective buying and selling of securities, and dissent at stockholders meetings, have become fixed on the American scene. The demand is growing for data that documents how effectively different companies meet human needs.[1]

Utility stockholders have the right to attend meetings and object to the investment of company money in reactors that have proven unreliable and expensive to repair. Investors in *any* company have a right to insist that the company take whatever steps are necessary to protect itself from bankruptcy in the event of a major nuclear power plant (or other nuclear facility) accident. Some nuclear power opponents have bought stock in order to have a voice. As another *New York Times* report put it: "All you need is one share."[2]

The Federal Communications Commission and Federal Trade Commission can also be enlisted to protest deceptive advertising by reactor manufacturers and utilities. A letter addressed "Dear Broadcaster," dated February 10, 1975, and signed by Representative Benjamin S. Rosenthal and thirty-one other members of Congress, opened with this statement:

> We are deeply concerned over the imbalance on the public airwaves created by many years and millions of dollars worth of government and industry promotion of nuclear power. Consequently, we are appealing to the nation's broadcasters to take affirmative action to assure that the public is fully and fairly exposed to all sides of this crucial issue.

Later in the letter reference was made to the Fairness Doctrine and to spot announcements prepared by the Public Media Center, which present the case against nuclear power plants.

Finally, it is important to remember that large religious and social service organizations can have significant financial—and political—clout. These organizations include churches, universities, foundations, as well as civic groups dedicated to the public interest. Many of these organizations—particularly the universities—own stock. Some of these groups also employ lobbyists who can be mobilized to work against nuclear power. In approaching such organizations, it can be pointed out that the safety (life and health) of present and future generations is a moral issue if there ever was one, that the preservation of a liveable environment has the deepest religious and moral implications, and that those who refuse to take a definite stand against these unprecedented risks are, in effect, supporting the proponents. As Edmund Burke said: "The only thing necessary for the triumph of evil [or error] is that good men [and women] do nothing."

Robert Rienow also put his finger on the issue in *American Problems Today* when he wrote:

> Great public problems cannot be ignored; they clamor for solution. And the citizen who sleeps is helping to decide, whether he wills it or not. The large associations, the groups, are lined up for and against an issue. They will force it to a verdict. To refuse to cast your influence one way or the other *is to stand for the stronger group.* As the Bible says, "He that is not with me is against me."[3]

Activating the Legislature

On the federal, state, and local levels, bills are always being introduced which have impact on the nuclear industry. Since the status of states' rights is questionable under the Atomic Energy Act, citizens and state officials should work together to clarify and reaffirm basic state's rights to protect the health and safety of the population within its borders. Pressure can be brought to gain votes in support of the antinuclear position. Sympathetic legislators can introduce antinuclear legislation—especially if they are provided with some background reading. One group that monitor federal legislation in this area is Critical Mass, the Ralph Nader affiliate in Washington (133 C St. N.W., Washington, D.C. 20036).

Several states have initiated far-reaching nuclear power restrictions. California's legislature passed through a couple of restrictive bills in 1976, in an attempt to undercut an antinuclear referendum. The vote failed, but further nuclear power development was banned until the California Energy Resources Conservation and Development Commission found that technology to reprocess and store radioactive waste existed. The Commission has not found this technology to exist, thus closing California to further nuclear development.

Voters in Missouri (1976) and Oregon (1978) voted to prohibit utilities from imposing charges for Construction Work In Progress, which adds plants under construction to the electric rate base and without which many utilities say they cannot afford nuclear power.[5] CWIP was also a major factor in the 1978 defeat of Governor Meldrim Thompson of New Hampshire, who strongly advocated nuclear power. Hawaii, meanwhile, restricted nuclear power in a constitutional convention.[6]

By far the most sweeping bill, however, was one enacted by referendum in Montana in 1978. Initiative 80, as the campaign was called, is a model for safe-energy advocates around the world. It establishes the citizens' right to say yes or no to each nuclear facility (excepting small medical and research, but including military, installations) in Montana and to establish nuclear safety and liability standards as follows:

1. 100 percent liability
2. Laboratory tests of all safety systems (including those involving transportation and other aspects of the fuel cycle)
3. Safe storage of radioactive materials
4. The cost of decommissioning and decontamination be paid for by the corporations responsible; a 30 percent bond must be posted to insure this
5. Evacuation and emergency medical aid plans must be published.

Ernie Krumm, one of the bill's organizers, explained some of the reasons why the initiative was successful, despite heavy lobbying from utility pressure groups. "The initiative was not a ban on nuclear power in Montana. Corporations building nuclear facilities in Montana will do so only if and when they demonstrate their respect for the rights of the voters in Montana by meeting the requirements stated in our "Bill of Rights."[4]

One critical area over which Congress has control is the loan subsidies for reactor exports. In a *New York Times Magazine* article, Leonard Ross makes a strong case.

The Export-Import Bank has made over $3 billion in loans and guarantees for reactor exports, at interest rates well below commercial levels. These favorable loans represent, in effect, a gift from American taxpayers to

American reactor manufacturers and foreign purchasers. Unilaterally ending this subsidy would do no damage to our economy. . . .

What is needed above all is a clearheaded understanding that the path to plutonium is paved with "demonstration" projects, billion-dollar trial runs, self-serving explanations that everyone else is doing it, and self-fulfilling prophecies that proliferation is inevitable. For the long run, we could place our bets on technologies less menacing than the plutonium-fed breeder—cleaner coal, cheaper solar power, enhanced conservation. For the near term, we could decide that the minimal economic benefits (if any) of plutonium recycling aren't remotely worth the security hazards.

We could reassess, along with the developing countries, whether their energy needs are best met by bargain-priced reactors or by monetary aid and accelerated development of non-nuclear technologies. We could avoid solutions to the reprocessing question—such as an exporters' cartel or a subsidized nuclear-fuel scheme—that would promote the uneconomic use of atomic power abroad. Most important, we could decide that the horrors of plutonium and proliferation deserve the same response that Franklin Delano Roosevelt gave to the Einstein letter that first described an atomic bomb: "This requires action."[5]

Citizens can demand from Congress action to insure safe energy through a moratorium on the building and licensing of nuclear power plants and the reinstatement of state and local community power to veto atomic installations; the repeal of the Price-Anderson Act; indemnification for nuclear investors, including retraining and provision for jobs in alternative power industries for displaced nuclear personnel; adequate funding of alternative sources of energy; and massive tax-reform measures enabling public interest groups to lobby legislators without jeopardizing their tax-exempt status.

On the local level, numerous cities and counties have acted on the transport of nuclear wastes, the levying of rate increases, and the municipalization of utilities.

On whatever level of government, representatives can be reminded that a pro nuclear stand will cost them votes in the next election.

Bringing It to the Courts

The judicial route is a slow and expensive way to oppose nuclear power. The earliest opposition to nuclear power manifested itself at the various public hearings required by law. Since the passage of the

National Environmental Policy Act court intervention continues to be a valuable antinuclear tool. Citizen groups can intervene not only on purely nuclear issues but on the economic impact—in other words, in hearings necessary for the frequent rate hikes. In this area, victories tend to come in small concessions, while the real battle is ignored. Friends of the Earth, Consolidated National Intervenors, and Union of Concerned Scientists are among the groups working in this area and other more widely reaching areas of safety and health.

Taking the Message to the Public Square

It was Albert Einstein who was quoted as saying the people must "take the message of atomic energy to the village squares." This is the sentiment and philosophy that characterizes the large public demonstrations that the antinuclear groups organize. This aspect of the movement traces back at least as far as Hiroshima and the Ban The Bomb crusade—which resulted in a limited test ban treaty. Newer groups, such as Mobilization for Survival, the various antinuclear alliances, and Ralph Nader's PIRG (Public Interest Research Group) network, work side by side with War Resisters League, SANE, Women's Strike for Peace, and Women's International League for Peace and Freedom. Such demonstrations are often tied in with visits to local legislators, union heads, and so on. There are also the civil disobedience groups who have been willing to go to jail, if necessary, to make their point.

In Europe, where antinuclear demonstrations have attracted 30,000 people on short notice, a reactor was halted by a large nonviolent permanent occupation of the site, in Weil, Germany. Over 28,000 people from several nations set up an anti-atom village and remained on the site until March 21, 1976—over a year later—when a court ordered a halt to construction. This inspired many actions in Europe—some successful, some not.

Taking cues from the Weil victory, several New Hampshire antinuclear groups coalesced to form the Clamshell Alliance; the name came from a cash crop which would be endangered by the Seabrook nuclear plant now under construction. The plant site, in addition to being an unusual ecosystem where birds and marine animals breed, happens to be close to an earthquake fault. It has been controversial since its 1968 announcement. After every legal avenue open to them could not stop construction from commencing, the newly formed group began planning an occupation on the Weil model. On August 1, 1976, just over a month after the NRC granted a construction permit, the brand new Clamshell Alliance held its first occupation. While six-hundred others rallied in support, eighteen people were arrested on the site.

The various Clamshell actions that followed received widespread press. New antinuclear alliances—named after all manner of local flora and fauna, as well as towns, rivers, and nuclear plants—sprang up in all parts of the country. Many of these have had their own occupations.

Since the accident at Three Mile Island, a more urbanized wing of the movement has appeared—one which concentrates heavily on the economic impact. These groups work on rate structures, employment opportunities, and actions at corporate headquarters; they are more concerned with the plight of poor people than the movement has been until now.

The antinuclear movement unites a wide range of people who might otherwise have very little in common, and as a result, many different tactics have been developed. The end goal of all of this activity is a conversion away from nuclear power and a move towards safe, renewable, decentralized energy sources. Also included in these goals are safe working conditions, a moratorium on nuclear power construction, repeal of the Price-Anderson Act, and strenuous efforts to safely store already accumulated radioactive wastes. United we can yet win the battle against nuclear myths, money, and madness.

Afterword:

An Analysis of the Accident at Three Mile Island

Richard E. Webb, Ph.D.

Some Consequences of the Nuclear "Experiment"

The accident hazards of nuclear power plants are much worse than the federal government, the promoter and regulator of nuclear energy, has so far acknowledged. The potential is great for a release of extremely large amounts of radioactive substances (in the form of vapors and dust) with harmful consequences to the public health and safety. The resulting radiation exposure and radioactive fallout could ruin agriculture over a land area the size of one-half of the United States east of the Mississippi River (500,000 square miles) for more than one hundred years. There would be at least one million victims of cancer, and a cloud of radiation could cause death by acute radiation sickness for a distance up to 75 miles downwind from a reactor and injury for many miles beyond. About 300,000 square miles would have to be permanently abandoned as a result of the release of plutonium radioactivity—a lung-cancer dust hazard—by a nuclear explo-

---

[Excerpted from an unpublished report entitled "The Urgency of Closing Down Nuclear Power Plants: An Analysis of the Three Mile Island Accident."]

sion accident in an advanced breeder reactor and conceivably in present-day power reactors as well.

In order to appreciate the gravity of these estimates, consider the Atomic Energy Commission's report *Thecretical Possibilities and Consequences of Major Accidents in Large Nuclear Power Plants* (WASH-740, March 1957), which calculates that a single reactor accident could cause agricultural restrictions over 150,000 square miles (the size of Illinois, Indiana, Ohio, and half of Pennsylvania) as a result of the release and fallout of strontium-90 radioactivity alone. Today's larger nuclear power plants, including the spent-fuel storage pool, can potentially release *500 times* more strontium 90 per plant than the reactor cited in the 1957 report. In other terms, a single reactor accident and the consequent eruption of the spent-fuel storage pool next to a reactor could disastrously contaminate the equivalent of more than 2,000 Lake Eries.

The number of different, potentially catastrophic reactor-accident possibilities is so large as to be virtually infinite, because of the great complexity of reactor plants and the limitless possibilities for sequences of equipment and component failures and human error. The instances of failures prior to the Three Mile Island accident have been relatively frequent.

The Three Mile Island (TMI) accident confirms what past experience indicated: that catastrophic accidents are indeed likely to occur and probably will occur in the near future, contrary to the federal government's assertions that the probability is remote. The reason for this expectation can be seen through the TMI accident, which was caused by *multiple* — at least five, in fact — equipment and component failures. In contrast, the safety systems of nuclear power plants are designed to control only those accidents caused by basically a *single* failure, except for a spontaneous rupture of the reactor vessel, for which nuclear plants are *not* designed to control.

Reactors are designed so that an accident caused by a single-system failure (except the reactor vessel), with everything else functioning properly, is *predicted* to be limited and therefore controllable. One of the problems of so-called reactor safety is that the official theoretical analyses of single-failure accident possibilities have yet to be verified experimentally; full-scale tests would be needed, and they are impractical. In the case of *multiple failures,* however, the reactor can explode, depending on the specific sequence of failures. The history of all past reactor accidents and near-accidents—including the TMI accident—shows clearly that whenever a reactor suffers a serious mishap, invariably *multiple* failures and/or human errors occur.

Although the NRC and other nuclear-energy promoters can claim that the risk of a catastrophic accident is low, and therefore acceptable, such claims are essentially speculative. The only sound, objective way to assess the safety of nuclear power is to consider *reactor experience*. Since this experience shows that reactor mishaps will probably involve multiple failures, we need to consider fully all multiple-failure accident possibilities, the course each could take, and the potential consequences. The NRC and its predecessor, the Atomic Energy Commission, have refused to investigate and analyze these possibilities, which explains why the public, Congress, and the utility operators of nuclear power plants are largely ignorant of them, and why the TMI operators were so confused as to what to do when the accident occurred.

That the TMI accident did not end in a core meltdown and a major public disaster is purely a stroke of luck. As with nearly all serious reactor-accident possibilities, neither analysis nor experiments existed prior to the accident to establish that it could have been controlled to prevent a reactor explosion and a catastrophic release of radioactivity. The TMI-2 reactor never was under control, nor is it now, for the accident is in effect an experiment for learning how nature will respond to a set of circumstances. We cannot predict with certainty the future course of this accident, since we know how nature behaves only by experiment and observation. The risk diminishes with time, as the heat source in the reactor—the radioactive fuel—decays. But the potential for an eruption still exists.

It is important to emphasize that the type of accident that occurred at TMI is among the least severe. Fortunately, the accident so far has been contained, although only after a considerable amount of jerry-rigging and luck. Thus the public still has an opportunity to investigate the hazards and take steps to close down nuclear power plants before we suffer an accident from which the country cannot recover.

The TMI-2 Accident of March 28, 1979

The accident at TMI-2 began with a failure of the main feedwater system, which caused the turbine steam-inlet valve to close. The steam relief valves then opened to discharge the steam from the steam generators into the atmosphere. With no makeup flow of water coming into the generators, they began to boil down, reducing their capacity to remove heat from the reactor coolant. As a consequence, the reactor coolant temperature and pressure rose rapidly. The power-operated relief valve on the pressurizer opened automatically to limit the rise in pressure, and the reactor control rods were automatically scrammed (inserted into the core) to stop the fission heat generation, which caused the core heat to fall to the decay heat level.

At this point, the situation was a "design-basis" accident—specifically, a complete failure of the main feedwater system—which was regarded by the NRC as a low-probability event. However, if it happens, the plant is designed to control the situation by scramming the reactor to stop the fission heating and activate the auxiliary feedwater pumps to continue the supply of water into the steam generators. The reactor designer's safety calculations predict that given the proper functioning of the relief valve and the auxiliary feedwater system, the reactor coolant system would be cooled down without incident by venting steam from the generator and eventually by the decay-heat removal system.

However, in the TMI-2 accident, the pressurizer relief valve stuck open, instead of closing when the reactor pressure fell. This failure caused a "loss-of-coolant" accident and depressurization of the coolant. Normally, the control-room operator could stop the relief flow by closing a separate valve that blocks off the relief valve. However, the position indicator of the relief valve also failed, indicating that the relief valve *was* closed. Presumably, the operators assumed that the loss of reactor pressure (coolant) was caused by a rupture in the coolant system, not by a stuck-open relief valve; so the block valve was not closed until two and a half hours later, when it was determined that the relief valve was indeed stuck open.

The reactor plant is equipped with an emergency coolant system (high-pressure injection, or HPI) to replenish the reactor coolant and thus control such a small loss-of-coolant situation. But the steam generator and the auxiliary feedwater system must be functioning properly to remove the bulk of the decay heat, since the heat removed from the reactor by the combination of coolant injection and steam flow through the pressurizer relief valve would be insufficient to discharge the core decay heat. In the TMI-2 accident the emergency coolant system automatically activated properly when the reactor coolant pressure dropped to 1,600 psi; however, because of the complete loss of feedwater and the boiling dry of the steam generators, this system was ineffective and could not prevent the coolant from overheating. As a consequence, the coolant pressure fell as its temperature rose, which together caused the coolant to boil and form steam within the reactor and its coolant piping.

Eight minutes into the accident, the auxiliary feedwater valves were opened and the steam generators began to fill with water, which restored the process of cooling the reactor coolant and thus removing the core decay heat. The reactor might then have been brought under control. However, the stuck-open relief valve was still discharging reactor coolant and thus venting the reactor pressure, which continued to produce steam bubbles in the reactor coolant system. (At fifteen minutes, the relief-valve discharge overloaded the drain tank, causing it to rupture. From then on, radioactive reactor coolant vented into the containment building.) The formation of steam bubbles in the reactor and its piping had evidently displaced coolant in the system, thus forcing it to flow into the pressurizer until it reached a high level. Evidently the operators assumed that the reactor coolant system was full of water (not steam), by the indication of the high water level in the pressurizer. So they stopped or throttled down the emergency coolant injection from time to time to prevent filling the pressurizer completely.

The steam formation in the reactor coolant also caused the coolant pumps to vibrate excessively. (They were designed to pump water, not steam.) This is the phenomenon of *cavitation*: steam or vacuum voids forming in the pump. In order to avoid damaging the pumps, the operators stopped them. Evidently the operators had assumed that the reactor core could be cooled by a natural convection circulation process: heated water emerging from the core rises because of convection and circulates through the steam generator, where it is cooled, and then circulates back through the reactor. Natural circulation cooling was previously analyzed by the reactor designers for a loss-of-coolant-pumping accident and was predicted to be adequate to cool the core. However, the steam bubbles in the reactor and coolant piping evidently increased the resistance to coolant flow and thereby interfered with natural circulation. (Previous natural circulation analyses may have assumed no loss-of-coolant situation or coolant depressurization.) In any event, the stoppage of the coolant pumps caused the coolant to overheat excessively, while the coolant flow through the core dropped drastically. With the pressurizer relief valve still open and venting pressure, steam formation in the reactor must have occurred more vigorously. During this period the fuel rods may have begun to overheat because of lack of cooling by water evaporation on their surfaces. The fuel heatup at that point may have been severe enough to cause zirconium-steam chemical reactions (zirconium oxidation), thus destroying or weakening the fuel cladding as well as producing large volumes of hydrogen gas.

An hour after the pumps were stopped, the operators determined that the relief valve was stuck open; they corrected the problem by closing the block valve ahead of the relief valve to stop the loss of coolant. With the pressure relief finally stopped, the reactor coolant pressure rapidly rose. The block valve was then opened for various periods over the next twelve hours or so and during attempts to depressurize the reactor to prevent excessive pressure.

At about 7.5 hours into the accident, the operators opened the pressurizer relief-valve block valve in an attempt to reduce the coolant pressure so that the decay-heat removal system could be activated to achieve a "cold shutdown" of the reactor. However, by reducing the pressure more steam formation must have occurred in the reactor, which perhaps caused more zirconium oxidation and thus more core damage and hydrogen production. During this and the preceding period, considerable quantities of hydrogen gas were probably blown out of the reactor system through the relief valve, accumulating in the containment building, where it evidently detonated. This hydrogen explosion momentarily pressurized the containment building to about half of its design pressure.

At 13.5 hours the operators aborted the attempt to activate the decay-heat removal system by closing the relief-valve block valve (to allow the pressure to rise), and by restarting one main coolant pump. The reason for this action is speculative. Perhaps because of the large amount of high-pressure hydrogen gas in the coolant system, the reactor pressure could not be reduced successfully to a low value. Perhaps the operators closed the relief valve to prevent the heavy release of radioactivity from the damaged fuel and confine it to the coolant system. Furthermore, the coolant must have been highly radioactive and therefore, if the decay-heat removal system were activated, the coolant would have been drawn out of the reactor containment, where the chances for a serious radiation leakage would have been greater. Perhaps the temperatures of the coolant above the core (which is instrumented to measure such temperatures at fifty-four places across its top) were reading high, and the operators judged that a high coolant flow was needed to cool the fuel. In any event, by restarting the reactor coolant pump the operators resumed core cooling by means of steam-generator operation. (Steam from the generator was vented to the turbine condenser.) The reactor then stabilized, but it was not under control. The problem remained: How to cool down the reactor?

At this point, a massive release of radioactivity from the damaged and crumbled reactor fuel rods into the coolant and reactor containment building had occurred. The NRC issued a memorandum contending that the release of radioactivity into the atmosphere was small: much less than the design-basis accident for the TMI-2 reactor allowed for. Specifically, the NRC estimates that about *one curie* of iodine radioactivity had escaped into the atmosphere, a fairly worrisome amount. (Iodine radioactivity is one of the most hazardous of all radioactive substances.) However, this author estimates that about *100 million curies* of iodine radioactivity had escaped the fuel, but it was fortunately confined to the coolant and the containment building. Therefore, the importance of the leak tightness of the containment building cannot be overemphasized.

During approximately the first five hours of the accident, highly radioactive reactor coolant escaped the containment building, because of a failure of the automatic containment isolation system to close the containment and the pipes that pass through its walls. But the coolant that escaped the containment flowed into and was confined to the auxiliary building. The air in the auxiliary building was ventilated and processed through charcoal filters, which presumably prevented a large escape of iodine into the atmosphere. (There is a question of whether and how much radioactivity may have been discharged to the Susquehanna River by the spillover of radioactive reactor coolant in the auxiliary building and drainage through the floor drains.) Had the containment building ruptured, the harmful consequences of the accident would have been severe. The potential for containment rupture was real, especially if fuel melting had not been averted. The possibility of fuel melting after the reactor had stabilized depended on the coolability of the core debris.

During the long periods when the coolant pressure was low enough to allow coolant to flash to steam, fuel overheating and heavy fuel damage must have occurred, particularly extensive zirconium-steam reactions, as evidenced by the large amount of hydrogen (formed by the reaction) that accumulated in the reactor vessel and in the containment building. This was indicated by excessive (off-scale) temperature levels measured at various places directly above the core by temperature sensors. These measured "core hot spots" were proof that the coolant-flow spaces between fuel rods in some regions of the core were partially blocked with zirconium dioxide debris, or that the fuel had simply crumbled as the result of zirconium fuel cladding being destroyed or oxidized by zirconium-steam reactions. Fuel crumbling would greatly impede coolant flow through the fuel and consequently would cause the coolant to pass slowly through the core. Hence it would reach much higher temperatures than coolant flowing normally through an array of fuel rods, which have no significant coolant flow blockage between them.

About a day and a half after the reactor was stabilized — by turning on a coolant pump and raising pressure — the NRC announced that a hydrogen gas bubble in the reactor vessel was interfering with the reactor cooldown effort. The danger, they said, was that when the coolant pressure is reduced, the bubble would expand into the core region of the reactor and thereby force water away from the core, causing the fuel to overheat again. Perhaps this was how the hydrogen gas bubble was first detected: by the early attempt to depressurize the reactor and observing possibly high above-core temperatures.

On April 1, the NRC released figures of the size of the hydrogen gas bubble in the core: about 1,500 cubic feet. (The bubble size was probably measured by injecting coolant under high pressure until the pressurizer filled; then the rate of pressure rise was measured as the hydrogen gas was compressed.)

On April 2, Dr. Earl Gulbransen, who is a leading authority on zirconium at the University of Pittsburgh, and myself calculated that the hydrogen gas volume was generated by oxidizing and therefore was destroying at least 25 percent of the zirconium cladding in the core. (Adding the amount of hydrogen vented into the containment building, the figure would be at least 35 percent.) Assuming that the zirconium-steam reaction occurred mostly in the upper half and center of the core, and away from its outer periphery, I concluded that the fuel in this region must be completely destroyed and crumbled as a result of the loss of the mechanical support normally provided by the cladding. Again, the existence of the core hot spots was direct evidence of such fuel crumbling. These hot spots may have continued to generate hydrogen and for days produced jets of steam at the top of the core, even though the core was submerged in water and the reactor was at a pressure high enough to suppress boiling. Based on his experiments and knowledge of zirconium, Dr. Gulbransen asserted that not all of the hydrogen generated would necessarily go to form the bubble—that some would be immediately absorbed by the remaining zirconium to form zirconium-hydride, a substance that would embrittle or weaken the remaining zirconium cladding. Thus there may be no upper limit as to how much of the fuel rod cladding has been destroyed or embrittled and no upper limit as to how much of the fuel has crumbled. If only one percent of the crumbled fuel had formed an uncoolable lump of fuel debris—that is, the material particles so compacted as to make the debris impermeable to coolant flow—the fuel in the lump would heat up to melting temperature, for the heat could not be conducted fast enough from the interior of the lump to its surface, where it could be carried off by surrounding coolant.

Once molten fuel is formed, the possibility exists for a "molten fuel-coolant interaction," which could result not only in a steam explosion but also in a runaway zirconium-steam reaction to produce more heat and hydrogen. Theoretically, a steam explosion caused by 1 percent of the core fuel (molten) or less can rupture the vessel, which in turn could cause a total loss of reactor coolant and then a total core meltdown and explosion. Because of the lack of prior core-destructive experiments, which are obviously impractical, and the impossibility of reliably calculating the behavior of a crumbled core— specifically, whether large, uncoolable fuel lumps could form to cause melting and explosions—because of mathematical limitations and limited physical data, the state of the core was completely unknown and therefore unpredictable. Anything could happen, no matter what action was taken, and nobody could predict the future course of the accident with certainty. But this was never recognized by the NRC, or if it was, they did not reveal it. Instead, the NRC pointed only to the hydrogen gas bubble, its potential interference with the cooldown process, and hence the need to eliminate the bubble, as if all that was needed to achieve a "cold shutdown" was carefully to vent the hydrogen from the reactor vessel and then depressurize the core.

Later that day, (April 2) NRC's Mr. Harold Denton, director of the TMI-2 accident control operation, announced that the hydrogen bubble had dissipated; shortly thereafter the NRC announced that the reactor operators could proceed to cool down the core by turning off the coolant pump and establishing natural convection cooling. However, the NRC gave no indication that the bubble was mostly dissipated by any venting process. On the contrary, Dr. Gilbransen asserts that much or most of the hydrogen could simply have been absorbed by the remaining zirconium in the core to form additional zirconium-hydride, thus consuming and embrittling the remaining fuel rod cladding. The hydrogen gas in the bubble dissolved in the coolant; the coolant, then flowing through the core, released the hydrogen to the fuel cladding.

In an NRC report attributed to Mr. Denton, the NRC stated that the dissipation of about 80 percent of the bubble was unexplainable, with the rest a result of venting, which supports Dr. Gulbransen's opinion that the remaining zirconium could have absorbed much or most of the hydrogen. Considering the size of the hydrogen bubble, the potential for hydrogen absorption by the core was such as to destroy all of the remaining zirconium fuel cladding in the core. Yet the NRC asserted that the dissipation of the hydrogen gas bubble was cause for "optimism" and thus implied that the crisis was over. On the contrary, however, the core was in an even weaker condition and more susceptible to crumbling and forming an uncoolable lump or lumps of fuel, making the situation even more dangerous because of the unknown state of the core.

The NRC originally feared that oxygen would evolve from water by action of the intense radiation (which decomposes the water into its parts, hydrogen and oxygen), and that the oxygen would then build up in the gas bubble to allow the hydrogen to burn explosively. The NRC subsequently concluded that the hydrogen from the gas bubble would dissolve in the water and chemically combine with the oxygen (reduce it back into water), thereby eliminating the oxygen. I assume the NRC's revised judgment on this hydrogen/oxygen explosion is correct. However, this correction gave the impression that the hydrogen gas bubble was not very serious after all. On the contrary, the hydrogen gas bubble was very serious, not only because it had threatened to uncover the core, but because the bubble indicated that a large fraction of the fuel had crumbled perhaps into an uncoolable mass.

With the hydrogen bubble dissipated, the question which then arose was how to cool down the reactor in a way which would minimize the risk of a core meltdown. The NRC announced that they were moving to effect a "cold shutdown" within about seven days. The NRC asserted that the cold shutdown state involved no further danger. (With the hydrogen bubble and the steam bubbles in the coolant system gone, the NRC evidently believed that the cooldown process could be accomplished.)

However, this plan was unwise, because without prior core-destructive experiments, it was impossible to establish whether or not such a drastic change in the mode of reactor cooling would trigger a compaction of fuel debris, thus causing a severe heatup and fuel melting. The process of changing over to a cold status would involve large reductions in coolant flow, possible "water hammer" shock waves (pressure surges) when switching to the decay-heat removal system, and other large effects, the consequences of which could be only conjectured.

Between April 2 and April 5 (and beyond), I warned the NRC and the Commonwealth of Pennsylvania authorities not to change the status of core cooling unless absolutely necessary. Caution required that the status quo be maintained. Because TMI-2 was in effect an experiment, and since the experimental results indicated that the reactor was stable under the one-pump/high pressure (1,000 psi) condition, the natural circulation mode of cooling should not be tested arbitrarily. Since the core decay heat *will* dissipate with time, the longer the reactor is cooled with the coolant pump, the lower the risk will be of a severe fuel heatup whenever the pump should have to be stopped. Thermal heat conduction theory predicts that after a one-year heat decay, the core cannot heat up to the fuel-melting temperature, even if the entire core is impermeable to coolant flow (except for possible momentary flashes of heat because of zirconium-steam reactions, which would still be dangerous, however).

It appears that the Pennsylvania officials were not aware of the possible crumbled state of the core, until they learned about it through discussions with me. On April 5 Governor Dick Thornburgh met with Harold Denton to discuss my information. In the end the NRC postponed switching off the main coolant pump; it let it run until April 27, when a pressurizer water-level instrument failed, forcing the operators to turn off the coolant pump and switch to the natural convection mode of cooling (removing heat through the steam generator). There was evidence of some degree of fuel heatup in the switch-over; but natural circulation so far has prevented a core meltdown and explosion.

Fortunately, the third and last pressurizer water-level gauge did not fail early in the accident when the other two gauges did. The gauge remained in operation for a month and allowed the main coolant pump to operate while the core heat source decayed by one-half to two-thirds of its early April level. If the pump had been stopped early in the accident, when the decay heat was two to three times or more greater, a core meltdown could have occurred. (The NRC's calculations confirm this possibility.)

Because of instrument limitations, the fuel heatup could not be measured, so we will never know the extent to which the fuel heated when the pump was stopped and therefore how close fuel melting and core explosion may have been approached. The so-called core hot-spot temperature readings, which the NRC has made public from time to time, were not temperature levels of the fuel but rather of the coolant emerging from the tops of the fuel-rod bundles or piles of fuel debris. The temperature sensors were located well above the actual fuel. Moreover, upon crumbling, the fuel would collapse and move down and further away from the temperature sensors. Thus it is not known to what extent coolant from colder regions of the core (regions that have experienced relatively little coolant blockage and have retained a relatively high permeability to coolant flow) was, or is, crossing over the top of the core and bathing a temperature sensor, thereby causing a false indication of the temperature of the coolant emerging from the fuel directly beneath the sensor. (The NRC acknowledges this possibility.)

The causes of the fuel heatup at TMI-2 may never be fully understood, because the information may be incomplete. For instance, the NRC says that the history of the operation of the high-pressure coolant injection system is "uncertain." Also, no temperature readings were taken of the above-core temperature sensors until four hours after the accident. Unless the time periods when the core overheated can be determined in order to establish just what caused the core to overheat, any corrective measures ordered by the NRC to try to reduce the probability of an accident recurrence will be unreliable.

Presently, the reactor is apparently stable. It still cannot be said to be under control, however, since the pressure must be kept high. The core could still deteriorate, by compaction or clogging coolant pores through fuel regions and causing the fuel to heat up, although this possibility would seem to diminish with time, as the heat-generation level decays. Again, the fuel could be heating up without necessarily being detected. All that can be said for certain is that it will take about one year from the start of the accident before the core decay heat will be reduced to such a low level that the possibility of sustained fuel melting would be excluded.

Lucky Circumstances of the TMI-2 Accident

Many lucky circumstances prevented the TMI-2 accident from ending in a core meltdown. Here is a partial list.

Lucky circumstance 1: *The core crumbled into a coolable geometry.* Perhaps pure chance was at work when the core overheated, crumbled, and formed into a configuration that was coolable. The core as easily could have collapsed into one or more uncoolable lumps. The random motion of the fuel-rod material upon crumbling could not be predicted. Neither could interdependent effects, such as how fuel crumbling would affect the heatup temperatures, which in turn would affect how finely the fuel fragments and compacts.

In addition, the zirconium-steam chemical reaction could simply have proceeded until it destroyed all of the fuel cladding and brought about a total core collapse and much greater hydrogen gas releases. Also, the heatup occurred at a rate slow enough to allow time to evaluate reactor data and to take action to control the reactor.

Lucky circumstance 2: *The operators aborted the switch-over to the decay-heat removal system at the right time.* At 7.5 hours into the accident, the reactor operators attempted to activate the decay-heat removal system by opening the pressurizer relief valve to depressurize the reactor coolant system. (The pressure had to be lowered below the maximum allowable pressure for the decay-heat removal system.) This caused coolant to flash to steam, and it probably contributed to core overheating. This condition lasted for six hours, when the operator closed the relief valve, thus repressurizing the reactor. The reactor core or a region of it may have been close to melting, so if the operators had waited any longer, even a half hour, the core may have melted or crumbled into a noncoolable configuration.

Also, it was fortunate that the operator did not activate the decay-heat removal system earlier, because the core heat-generation level would have been much greater and the core heatup during depressurization more severe.

Lucky circumstance 3: *The hydrogen gas bubble dissipated.* The dissipation of the hydrogen gas bubble in the reactor was largely unaccounted for, raising the possibility that the remaining unoxidized fuel cladding had absorbed the hydrogen (to form zirconium hydride). If this hydrogen absorption process had not occurred, the presence of the bubble may have been prolonged, adding more core cooling problems. In addition, the danger of an explosion in the containment building would have been greater because more hydrogen would have been vented into its atmosphere. Of course, the formation of zirconium hydrides would make the core more susceptible to collapse.

Lucky circumstance 4: *The accident occurred after only three months of reactor operation.* Fuel rods undergo great changes as a result of atomic fissioning. The zirconium fuel cladding, which is mechanically ductile when new, becomes more brittle through neutron irradiation and corrosion. Also, the uranium oxide fuel material inside the rod becomes swollen as a result of the buildup of fission products. These factors mean that the core would more likely have totally crumbled, thus creating an uncoolable heap of fuel debris if the reactor had been older. Furthermore, the core-decay heat levels would have been greater, because of the greater buildup of radioactivity in the core. Specifically, the decay-heat level would have been 76 percent greater than the level that actually existed when the switch-over to natural circulation occurred (30.5 days into the accident). Or, the heat level would have been a significant 15 percent higher during the period of the core heatup in the initial phase of the accident.

Over time, the reactor vessel itself becomes embrittled by neutron irradiation, which requires strong limitations on the coolant pressures whenever the temperature of the vessel walls falls below its operating level. During the TMI-2 accident, the situation arose where the coolant pressure was high (2,100 psi), and yet the temperature of the coolant entering the reactor vessel was low (150° F.), which means that a portion of the vessel must have been cold while the vessel was under high pressure. In general, all parts of the reactor coolant system weaken with use. Had the accident occurred later, the mechanical strength of the materials throughout the system (vessel, pressure housings, closure heads, and so forth) would have been less, and consequently the various loadings and stresses produced by the temperature changes and pressures of the accident may have caused ruptures from the brittle fractures.

Lucky circumstance 5: *The reactor was scrammed.* The accident data shows that the reactor automatically scrammed at about nine to twelve seconds into the accident; that is, the fission power was stopped, whereupon the core heat-generation rate dropped promptly to the decay-heat level (from 100 percent reactor power to about 5 percent). The scram involves the rapid insertion of about sixty control rods to quash the reaction. However, if only two or three adjacent control rods had failed to insert, then the atomic fissioning could have continued in a region of the core, thus melting the fuel. Had a scram failure occurred, the reactor power would have remained high, and a rapid core meltdown and reactor explosion surely would have occurred.

Lucky circumstance 6: *The pressure relief valve stuck open instead of functioning normally.* The loss of feedwater that started the TMI-2 accident caused the reactor coolant to overheat and the coolant pressure to rise. To prevent coolant overpressure, the pressurizer relief valve popped open to vent steam. Because the relief valve stuck open, however, it caused the coolant pressure to decrease to levels at which the coolant boiled. The steam thus formed could have at that point deprived the core of adequate coolant, thus starting fuel heatup. Certainly, the steam caused cavitation of the coolant pumps, which required the pumps to be stopped. Stopping the coolant pumps then either caused the fuel to begin to heat up or worsen the heatup. Therefore, that the relief valve stuck open was a factor in producing the accident.

What would have happened if the relief valve had functioned properly later in the accident, if it had closed when the pressure became stable, instead of allowing the pressure to continue to fall? Because the accident involved a complete loss of feedwater, it therefore would still have been worse than the design-basis feedwater accidents. The coolant temperature and pressure would have remained high, and the coolant loss from the reactor system would have been minimum. This means that the pressurizer would presumably have filled with water as a result of the thermal expansion of the coolant upon heatup. Evidently, filling the pressurizer is dangerous, but the dangers have not been explained in the safety literature.

Could severe water-hammer shock waves have occurred had the relief valve functioned properly? Such pressure loadings could cause a rupture in the reactor coolant system and thus a loss-of-coolant accident. (The emergency core-cooling system is not designed to control such an accident.) Or, if the relief valve stuck open after the pressurizer filled, would the steam formation in the reactor upon the drop in pressure have been worse than the TMI-2 accident? The coolant would be hotter. Also, other things could have activated and then could have malfunctioned, such as the safety valves, which vent coolant at a higher rate than the power-operated relief valve.

The NRC has not analyzed a complete loss-of-feedwater accident, where there is no auxiliary feedwater but where the relief valve functions properly at least until the pressurizer fills. The NRC discusses it only qualitatively in its recent *Staff Report on the Generic Assessment of Feedwater Transients in Pressurized Water Reactors Designed by the Babcock and Wilcox Company* (NUREG-0560, May 1979). The NRC asserts that no core damage would occur in the first twenty minutes, provided that coolant is injected into the reactor by manually activating the high-pressure injection system (HPI) to reduce the temperature of the reactor coolant, and provided two of four main coolant pumps are stopped to reduce the heat generated by the pumps.

Since the auxiliary feedwater system was activated within twenty minutes in the TMI-2 accident, the NRC implies that the accident would not have been serious if the relief valve had functioned properly. However, there is no reason to believe that the HPI system would have been activated manually, since the TMI operators turned off the HPI system after it had automatically started up because the pressurizer water-level gauges indicated a high water level. (The HPI had activated automatically when the pressure dropped to 1,600 psi—the automatic activation point.) Had the relief valve functioned properly during the TMI-2 accident the operators would most probably have acted under the same assumptions, and they would *not* have taken the actions which the NRC's *Feedwater Transients* report asserts would be needed to control the accident.

Reactors and the accident processes are too complex to admit reliable guesswork or estimates based on intuition and simple logic arguments. If the TMI-2 core could have still suffered damage had the relief valve functioned properly, as the NRC's NUREG-0560 report implies for the circumstances and operator actions and inactions of the TMI-2 accident, then there would be no way to prove that the accident would not have ended in catastrophe, except by conducting a full-scale experiment. All we know for certain is that the set of equipment failures that occurred at TMI-2 produced a serious, but not catastrophic, accident.

Lucky circumstance 7: *The auxiliary feedwater system failed.* It may have been lucky that the auxiliary feedwater system *failed* to function properly when the relief valve stuck open. NUREG-0560 points out that the nonfailure possibility "represents the most severe case of depressurization" of the coolant system. This is because the steam generators would always contain water (feedwater) after the fission power stops, and so the coolant would be cooled down rapidly. The NRC report states that the coolant may boil in the reactor. Since coolant boiling in the reactor system caused the core to overheat, the pumps to cavitate, and so on, this accident could have ended in core damage and therefore disaster.

Lucky circumstance 8: *The hydrogen explosion did not cause additional failures.* At about ten hours into the accident, a hydrogen explosion evidently occurred, which caused a pressure blast of about one-half of the containment design pressure. Although this explosion did not rupture the containment building, it was fairly strong. But the explosion could have been more powerful. The chance detonation could have occurred later, thereby allowing more time for the venting of hydrogen from the reactor and thus a greater accumulation of hydrogen in the containment atmosphere to fuel the explosion. Also, the core heatup could have generated more hydrogen, had the sequence and timing of the operator actions been different. Moreover, it is not known just *how* the hydrogen was concentrated. Possibly the hydrogen could have accumulated and concentrated in such a region of the containment building as to cause a rupture.

The explosion may have come close to rupturing one or more control-rod drive mechanisms attached to the top of the reactors, by the effect of shock and blast waves. These mechanisms are not designed to withstand hydrogen explosions — nor is any other reactor equipment for that matter. If one or more motor tubes had ruptured, the associated control rods would have been ejected from the reactor core. The strong reactor pressure would have expelled the control-rod drive shaft that extends into the motor tube and hence the control rod connected to it. It is conceivable that the ejection of several control rods could have caused a nuclear-fission power excursion, which would have greatly worsened the accident and potentially could have caused an instant reactor explosion and consequent containment rupture. Eventually during the accident, a possible recurrence of nuclear fissioning was prevented by injecting boron into the reactor coolant. Boron acts to suppress fission. But whether the boron concentration was high enough when the hydrogen explosion occurred, such that it would have prevented a fission power excursion had several motor tubes ruptured, is a question that must be explored.

Even if a fission power excursion was not possible when the explosion occurred, a ruptured motor tube would probably have caused a core meltdown by venting the reactor coolant and depressurizing the reactor; the core debris would continue to overheat by steam formation in the core. Only by repressurizing the reactor, by closing the relief valve, and starting up a coolant pump, did the TMI-2 reactor finally stabilize. That is, when the reactor was undergoing depressurization, the core severely overheated. Also, a ruptured motor tube would have resulted in a massive, rapid release of hydrogen into the containment from the reactor (from the hydrogen bubble), which could have then caused a more severe, second hydrogen explosion and a containment rupture.

Although the hydrogen explosion apparently did not rupture any control-rod drive-mechanism motor tubes, certain conditions could have increased the likelihood of a rupture. First, the hydrogen explosion could have been more powerful because of a greater accumulation of hydrogen in the containment before detonation. Second, the hydrogen explosion was the result, certainly, of a haphazard formation and detonation of hydrogen gas outside the reactor vessel. Conceivably, the explosion could have produced pressure waves, imposing greater mechanical forces on the motor tubes. Third, the hydrogen explosion occurred when the reactor pressure was at a relatively low (500-psi) level. Shortly after the explosion, but not because of it, the reactor coolant pressure was raised to 2,300 psi. Had the hydrogen detonated when the reactor pressure was at this higher level, the steel walls of the motor tubes would have been under much greater pressure and indeed may have ruptured. Finally, other parts of the reactor system could have ruptured under a hydrogen explosion: Pipes, for example, were not designed to withstand explosions.

Lucky circumstance 9: *The operators were able to fix the failed auxiliary feedwater valves in eight minutes.* Had more time elapsed before the feedwater flow was restored (by opening the two valves), the reactor coolant and the core would have reached even higher temperatures. Potentially, the hottest fuel could have increased another 1,000° F. in five minutes, so that if additional time had elapsed, there could have been a core meltdown, either during the initial phase of the accident or later, when the operators were forced to switch off the coolant pumps.

Lucky circumstance 10: *The third and last water-level gauge on the pressurizer functioned for one month instead of failing early in the accident when the other two redundant gauges failed.* The TMI-2 pressurizer was equipped with three independent water-level gauges. At least one gauge was needed to ensure control of the coolant level in the reactor system and to ensure a sufficient-size steam bubble in the pressurizer to avoid severe pressure rises while one of the main reactor-coolant-recirculation pumps operated. However, two of the three gauges failed early in the accident, presumably because of excessive radiation exposure. The NRC (Dr. Roger Mattson) said that the gauges were designed to withstand a radiation exposure level of 100,000 rads. However, the rate of radiation inside the containment building quickly reached 30,000 rads per hour in the first day or two of the accident. Hence, the gauges would be expected to fail 3.3 hours after such high radiation levels. Sure enough, two gauges failed after about six hours under the high radiation level, according to Dr. Matt-son. Had the third gauge also failed, the accident would have taken a much worse course, perhaps resulting in a core meltdown. Without the pressurizer water-level control, the powerful main coolant pump would have had to have been stopped, and natural circulation cooling attempted. Natural circulation coolant flow would hardly have been adequate to prevent a runaway fuel heatup, because of the two-to-three-fold greater core decay-heat levels. Attempting to cool the core by the decay-heat removal system would have required reducing the coolant pressure. But this very procedure caused the core to overheat during the attempt at 7.5 hours to activate the system. As it was, the remaining gauge miraculously continued to function for a month, after the gauge was exposed to 13.5 million rads of radiation, or 136 times its design limit, allowing the coolant pump to operate for that critical period of core cooling.

Lucky circumstance 11: *The crumbled fuel did not form a critical mass of fuel to start a runaway reaction.* The reaction was terminated 9 to 12 seconds after the accident by the rapid insertion of the control rods into the core. But the control rod material evidently melted dur-ing the core heatup; enough control material *could* have migrated away from a region of the crumbled fuel to cause a fission reaction in the pile of fuel granules. To prevent such a possibility, the TMI-2 operators injected additional boric-acid neutron absorber into the coolant. Normally, the coolant contains boric acid to help control the atomic fission reaction in the core during power operations. At the be-ginning of the accident, the boron concentration in the coolant was 1,030 parts per million. The operators eventually raised the boron con-centration to 3,000 ppm., to prevent fissioning. Based on fission-theory calculations, the NRC concluded that the crumbled core could possibly cause fissioning if the boron concentration were fewer than 3,000 ppm.

However, the concentration of boron was low for an indetermi-nate time. It is possible that the core debris came very close to a fis-sion reaction during the initial phase of the accident. If so, it would have been a simple stroke of luck that the core debris settled into a nonfissioning configuration. The intense heat generation of a fission reaction could potentially cause a core meltdown.

The NRC report mentions that the damaged reactor condition

precluded the neutron detectors in and outside the reactor from indicating if a fission reaction was forming. Therefore, there is no way to know how close the crumbled core had come to a fission runaway until the core debris is examined, if the core *can* be chemically examined for concentration of residual control-rod material. Furthermore, there could have been no prior warning had the fissioning occurred.

It should also be noted that the theory used by the NRC in calculating that 3,000 ppm. of boron is sufficient to preclude fission in a fuel debris pile may not have been adequately verified. One would have had to experiment with *fuel debris,* not fuel rods or plates, as is customary in fission experiments. The complexities of "neutron transport" theory would seem to dictate the use of fuel debris in experiments, but it is doubtful that such experiments have ever been performed. Hence, a fission runaway could conceivably still occur.

In addition, if the core heatup had been more severe, the core debris may have formed a configuration that would have undergone a runaway fission reaction. Again, such possibilities cannot be determined, except by full-scale reactor experiments or by experiencing the accidents.

Lucky circumstance 12: *Nothing else failed.* Considering the many failures and the large thermal changes that occurred, the possibilities are innumerable.

Lucky circumstance 13: *A combination of worse things did not occur.* In addition to the many separate possibilities that could have ended in catastrophe, possibly two or more of such separate possibilities could have occurred, which obviously would have increased the severity of the accident.

We can conclude from the preceding analysis that the prime ingredient in the Three Mile Island accident was neither intelligence nor design, but simply luck.

Nuclear Semantics

Several comments on the TMI-2 accident issued by the NRC, the Department of Energy, and industry officials promoted popular misconceptions of the seriousness of the accident and a false sense of security about the state of the TMI-2 reactor—and about nuclear power plants in general. These deceptive comments pertain to the NRC's characterization of the efforts to shut down the reactor safely as a "cooldown" and the reactor condition as "stable" during the critical hours and days of the accident and the assertion that "the system worked."

First, the so-called cooldown is inaccurate. It was never certain whether the fuel was cooling down or in fact heating up. When the switch-over to natural circulation occurred, the fuel did heat up.

Similarly, it would have been more accurate to characterize the reactor as *"apparently* stable." Because the status of the fuel was uncertain, "stable" is inaccurate.

The assertion that "the systems worked" is also incorrect. *None* of the reactor plant safety systems was designed to control the accident that happened. Rather, the system was amenable to being controlled with considerable jerry-rigging. For example, hydrogen burners had to be transported to the site and installed in high radiation areas to prevent further hydrogen explosions in the containment building. Methods had to be devised to determine the size of the hydrogen gas bubble in the reactor and to vent the bubble. Huge diesel-powered electric generators, located and transported to the site, provided a backup supply of nine megawatts of electricity to power reactor-coolant pumps in the event of a failure of electrical power from the off-site, electric-power grid. A backup source had to be constructed by connecting a power line into a separate power grid. A backup reactor-pressure-control system had to be built. An improved, leakproof, decay-heat removal system had to be fabricated and shipped to the site. Special heat exchangers (water coolers) had to be ordered, and steam-generator systems had to be modified to remove heat without creating steam, and so on. The systems did *not* work.

The TMI accident demonstrates that the NRC has no plans for coping with serious accidents. Rather, the NRC's philosophy is to assess an accident and then develop at that time a means, if possible, to effect a safe condition. Consider the design-basis loss-of-coolant accident, an accident considered to be "credible" by the NRC. The safety analysis report is mute about how the reactor could be cooled reliably for a long term and how the reactor containment could ever be opened and entered to effect repairs. The fact is, no plans exist for repairing a reactor that has suffered a simple pipe rupture, large or small.

So far, the NRC has issued two reports: *Staff Report on the Gen-*

eric Assessment of Feedwater Transients in Pressurized Water Reactor Designed by Babcock and Wilcox (NUREG-0560) and Evaluation of Long-Term Post-Accident Core Cooling of Three Mile Island Unit 2 (NUREG-0557). (Here again are subtle semantics. By post-accident, the NRC suggests that the accident is over. On the contrary, the accident is still going on, for the reactor is still not under control; rather, it is apparently stable. The operators are not able to reduce the system pressure, for the core could conceivably still flare up if the pressure were reduced.) These reports provide much useful information, but they are far from adequate.

A full analysis of the TMI-2 accident is necessary if we are to learn the full dangers that existed (and still exist). However, we must not become too preoccupied with evaluating the TMI-2 accident, because so far as future safety is concerned, the report will be of no use. There are virtually an infinite number of more severe reactor-accident possibilities for which the TMI-2 report would not be applicable. Nevertheless, when the NRC eventually issues its complete report on TMI-2, it will be scrutinized with great care.

Conclusion

Nuclear power plants are unsafe. We must close them down immediately and dismantle them. Of course, doing so would have profound implications for the present, highly industrialized way of life in America and its economic order. However, I believe that these would be to the good, and that a non-nuclear way of life is viable—with a transition from heavy oil and coal usage to solar-based energy (solar rays, wood, wind, and minimal fossil usage)—requiring about seventy-five years to make the transition.

Who should decide the issue of nuclear energy? The Constitution and historical evidence of the intentions of the makers of the Constitution run completely contrary to the fairly recent assumptions by Congress that they have the power to promote industry by expenditures. The federal laws for the promotion and regulation of nuclear power are simply unconstitutional—the people have not granted to Congress the power to promote nuclear energy.*

---

*I am opposed to civil disobedience and urge, instead, public education, lawful petitioning of government, and resolution of the nuclear issue through constitutional processes, as discussed in my book The Accident Hazards of Nuclear Power Plants (The University of Massachusetts Press, 1976).

that perilous situation. It is only before an accident occurs that steps can be taken to avoid such situations. In view of the virtually infinite number of accident possibilities, the only step that can be taken to ensure against catastrophic accidents while society reviews nuclear energy is to shut down reactors and place a heavy security guard to maintain the needed continuous operation of the decay-heat cooling systems.

References

The references for the description of the TMI accident events are:

1. The NRC's "Preliminary Notification of Occurrences" and "Board Notifications" issued following the start of the accident.
2. "Safety Analysis Report for Transition to Natural Circulation," issued by Metropolitan Edison, April 12, 1979.
3. *Evaluation of Long Term Post-Accident Core Cooling of Three Mile Island Unit 2,* NRC Staff Report, NUREG-0557, May 1974.
4. *Staff Report on the Generic Assessment of Feedwater Transients in Pressurized Water Reactors Designed by Babcock and Wilcox Company,* NUREG-0560, May 1979, NRC.
5. *Investigation into the March 28, 1979, Three Mile Island Accident by the Office of Inspection and Enforcement,* NUREG-0600, August, 1979, NRC.
   See also, Richard E. Webb, *The Accident Hazards of Nuclear Power Plants,* University of Massachusetts Press, 1976.

Glossary

AEC Atomic Energy Commission, responsible for promoting and regulating atomic energy through 1974.

Chain Reaction The process by which atomic fission becomes self-sustaining; a neutron fired at the nucleus of an atom will release other neutrons, which in turn will hit other nuclei.

Critical Mass The point at which the amount of material being fissioned is large enough to start, and continue, a chain reaction.

Curie A standard unit of radioactivity describing the number of atomic fissions per second and equal to that of one gram of radium. Curies describe the number of emissions from a radioactive source, whereas roentgens describe the amount of radiation energy an object absorbs.

Fission Splitting atoms. This can occur naturally, or when an atom's nucleus is bombarded by fast-moving neutrons.

Fissionable Material that can be fissioned; also known as fissile.

Fission Products Nuclear waste formed by the splitting of a large atom into smaller ones, as well as the products formed by the radioactive decay of the original products into smaller atoms.

Half Life Decay in a radioactive element is described in terms of its half life, the time it will take for half of its atoms to disintegrate through fission. The isotope ruthenium 106, for example, has a half life of one year, meaning that one pound of it will be reduced to one-half pound one year later and one-quarter pound a year after that.

Megawatt A measure of electrical generating capacity. A kilowatt is a thousand watts, while a megawatt is a thousand kilowatts, or a million watts. A kilowatt-hour is the amount of energy consumed by a thousand watts burning for one hour.

NRC Nuclear Regulatory Commission, successor to the AEC, in theory charged only with regulation, although in practice almost as concerned with promotion as the AEC had been.

Picocurie A trillionth of a curie (1/1,000,000,000,000).

Rad A measure of an absorbed dose of radiation. Millirad is one-thousandth of a rad.

Reactor-Year A measure of our total experience with nuclear reactors. The total number of years each reactor (including those no longer in use) has been operating, added together, equal the total number of reactor-years.

Rem Roentgen Equivalent Man, another measure of radioactivity. This measures the potential effect on human beings. For some elements a rad and a rem are equivalent. For weaker elements the amount of rems a person received would be less than the number of rads.

Roentgen A measure of the exposure to X rays or gamma rays, named for William Roentgen, who discovered X rays. Roentgens are a measure of the actual energy absorbed in tissue. Gamma rays are a more penetrating radiation than alpha or beta radiation, but the latter two are more concentrated, and can do more damage once inside the body.

Scram The immediate insertion of all control rods into the reactor's guts, resulting in an emergency shutdown of the plant.

Notes

The following list omits those references whose sources have already been fully identified in the text of this book.

Introduction: Lessons from Three Mile Island

1. Hannes Alfven, "Energy and Environment," *Bulletin of the Atomic Scientists,* May 1972.

2. "A Nuclear Nightmare," *Time,* April 9, 1979.

3. *Washington Post,* April 14, 1979.

4. Charles Mohr, *New York Times.* Reprinted, *St. Petersburg Times,* April 13, 1979.

5. *Atlantic City Press,* April 14, 1979.

6. Ibid.

7. *Atlantic City Press,* August 3, 1979.

8. *Washington Post,* April 15, 1979.

9. *Critical Mass Journal,* December 1978.

10. *Critical Mass Journal,* July 1979.

11. *Nucleus,* May 1979.

12. *Providence Journal,* November 2, 1977.

13. Anna Gyorgy and Friends, *No Nukes* (Boston: South End Press, 1979), p. 106.

14. Richard E. Webb, *Accident Hazards of Nuclear Power Plants* (Amherst, Mass.: University of Massachusetts Press, 1976), p. 195.

15. Gyorgy, *No Nukes,* p. 120.

16. Ibid., p. 357.

17. Ibid., p. 112.

18. Except for Three Mile Island, this list was based on data from Gyorgy, *No Nukes,* pp. 117–18; and Webb, *Accident Hazards,* pp. 187–201.

1. The Goose That Laid the Golden Egg

1. The account of the Fermi Accident is based on the Atomic Energy Commission, *Preliminary Report on Fuel Damage in Fermi Reactor,* AEC Docket No. 50-16, October 11, 1966; J. G. Duffy, W. H. Jens, J. G. Feldes, K. P. Johnson, and W. J. McCarthy Jr., "Investigation of the Fuel Melting Accident at the Enrico Fermi Atomic Power Plant," paper presented at the national topical meeting of the American Nuclear Society, San Francisco, April 1967; W. J. McCarthy Jr., additional remarks made while delivering the latter paper; and Sheldon Novick, "Breeding Nuclear Power," *Scientist and Citizen Magazine,* June/July 1967.

2. Robert E. Beardsley, *The Natural Philosopher,* vol. 1 (Blaisdell Publishing, 1963).

3. Anna Gyorgy and Friends, *No Nukes* (Boston: South End Press, 1979), p. 338.

4. John J. Berger, *The Unviable Option* (New York: Dell, 1977), p. 132.

5. Ibid.

6. Saul Friedman, "The Enrico Fermi Power Plant," *Detroit Free Press,* July 17, 1966.

7. This and subsequent references to the Fermi legal action are taken from Brief for Petitioners, United States Court of Appeals, No. 15271 (decision rendered June 10, 1960); Brief for Respondents, No. 315 and No. 454, Supreme Court of the United States, October Term, 1960; Brief of Adolph J. Ackerman, Amicus Curiae, No. 315 and No. 454, Supreme Court of the United States, October Term, 1960.

8. Testimony before the House Appropriations Subcommittee, June 29, 1966.

9. Engineering Research Institute, The University of Michigan, *A Report on the Possible Effects on the Surrounding Population of an Assumed Release of Fission Products into the Atmosphere from a 300 Megawatt Nuclear Reactor Located at Lagoona Beach.* (Ann Arbor: University of Michigan, July 1957).

10. Richard E. Webb, *Accident Hazards of Nuclear Power Plants* (Amherst, Mass.: University of Massachusetts Press, 1976), pp. 3–4.

11. *Manchester Guardian,* March 19, 1978.

12. U.S. Naval Ordnance Laboratory, NAVORD Report 5747, October 7, 1957, pp. i–ii.

13. *The National Observer,* November 6, 1967.

14. Sheldon Novick, "Continuing the Fermi Story," *Scientist and Citizen Magazine,* November–December 1967, pp. 224–25.

15. *Nucleonics,* December 1966.

16. John Fuller, *We Almost Lost Detroit* (New York: Reader's Digest Press, 1975), pp. 223–28.

2. Those Who Favor Fire

1. Historical and factual background—though not necessarily the interpretations thereof—is taken, except where noted, from John F. Hogerton, "The Arrival of Nuclear Power," *Scientific American,* February 1968, pp. 21-31.

2. Statement before the New York City Council, June 14, 1963.

3. Barry Commoner, "Fallout and Water Pollution—Parallel Cases," *Scientist and Citizen Magazine,* December 1964.

4. References to Atomic Energy Act and amendments from *Atomic Energy Act of 1946 and Amendments,* Gilman G. Udell, comp., (Washington: U.S. Gov't Printing Office, 1966), pp. 71-2460.

5. T. E. Murray, *War and Peace* (Cleveland: World Publishing, 1960).

6. David E. Lilienthal, "When the Atom Moves Next Door," *McCall's,* October 1963.

7. Saul Friedman, "The Enrico Fermi Power Plant," *Detroit Free Press,* July 17, 1966.

8. James W. Kuhn, *Scientific and Managerial Manpower in Nuclear Industry* (New York: Columbia University Press, 1966), p. 114.

9. *St. Petersburg Times,* April 20, 1979.

10. *Congressional Record,* July 16, 1963.

11. Hogerton, "Arrival of Nuclear Power." *Scientific American,* February 1968.

12. Kuhn, *Scientific and Managerial Manpower,* p. 115.

13. See references in chapter 13 for Price-Anderson Act sources.

14. John J. Berger, *The Unviable Option* (New York: Dell, 1977), p. 146.

15. Kuhn, *Scientific and Managerial Manpower,* p. 115.

16. Atomic Energy Commission, *Civilian Nuclear Power: A Report to the President,* November 20, 1962, p. 4.

17. David E. Lilienthal, *Change, Hope, and the Bomb* (Princeton, N.J.: Princeton University Press, 1963), p. 96.

18. Data on subsidies from Drew Pearson, "Washington Merry Go-Round," *Washington Post,* July 9, 1963.

19. *Congressional Record,* July 16, 1963.

3. Thresholds of Agony

1. Ernest Sternglass. Speech given at a rally against nuclear power at Jamesport, N.Y. (Courtesy of WBAI News Department).

2. *Federal Register,* June 9, 1971, p. 36. In Peter Faulkner, *The Silent Bomb* (New York: Vintage, 1977), p. 246.

3. *The Hazards to Man of Nuclear and Allied Radiations* (London: Medical Research Council, 1956).

4. Helen Caldicott, *Nuclear Madness* (Brookline, Mass.: Autumn Press, 1977). p. 51.

5. James F. Crow, "Radiation and Future Generations." In John M. Fowler, ed., *Fallout* (New York: Basic, 1960).

6. W. C. Hueper, "Recent Developments in Environmental Cancer," *American Medical Association Archives of Pathology,* 1954, pp. 58475–523.

7. Dr. Martell's work was examined and quoted in great depth in chapter 5 of McKinley C. Olson, *Unacceptable Risk* (New York: Bantam, 1976). Though Martell is mentioned in other sources (Nader and Gofman, to name two), the information here is taken almost entirely from Olson, pp. 87–142.

8. Ibid., p. 125.

9. Ibid., pp. 89–92.

10. Ibid., p. 91.

11. Rachel Carson, *Silent Spring* (Boston: Houghton Mifflin, 1962).

12. Robert S. Stone, "Maximum Permissible Exposure Standards." In C. R. McCullough, ed., *Safety Aspects of Nuclear Reactors,* vol. 13 (New York: Van Nostrand, 1957).

13. Walter R. Guild, "Biological Effects of Radiation." In Fowler, *Fallout,* p. 84.

14. Isaac Asimov and Theodosius Dobzhansky, *The Genetic Effects of Radiation,* U.S. AEC Division of Technical Information, Library of Congress Catalog Card Number 66-62747, 1966, pp. 26–38.

4. Nuclear Roulette

1. T. J. Thompson and J. G. Beckerley, eds., *The Technology of Nuclear Reactor Safety,* vol. 1 (Cambridge, Mass.: M.I.T. Press, 1964), p. 5.

2. C. R. McCullough, N. M. Mills, and Edward Teller, "The Safety of Nuclear Reactors." In McCullough, ed., *Safety Aspects of Nuclear Reactors* (New York: Van Nostrand, 1957), p. 140.

3. Atomic Energy Commission, *Small Nuclear Power Plants, COO-284,* vol. 1 (Washington, D.C.: Atomic Energy Commission, Division of Technical Information), pp. 24–26.

4. Robert E. Beardsley, *The Natural Philosopher,* vol. 1 (Blaisdell Publishing, 1963), p. 35.

5. Thompson and Beckerley, *Technology of Nuclear Reactor Safety,* p. 35.

6. Clifford K. Beck, "Engineering Out the Distance Factor," *Atomic Energy Law Journal* 5, no. 4, (Winter, 1963).

7. U.S. Congress, Joint Committee on Atomic Energy, *Atomic Energy Commission Authorizing Legislation, Fiscal Year 1968,* Part 2, March 14 and 15, 1967, p. 771. (Abbreviated from hereon as *Auth. Leg. 1968.*).

8. *Nucleonics Week* 8, no. 15, April 13, 1967.

9. *New York Times,* April 25, 1965.

10. *Auth. Leg. 1968,* p. 121.

11. U.S. Department of Commerce, Tornado Statistics, ESSA/PI 6600 29.

12. *New York Times,* April 25, 1965.

13. "Newsmakers," November 21, 1965, program by Columbia Broadcasting System.

14. John J. Berger, *The Unviable Option* (New York: Dell, 1977), p. 45.

15. Ralph Nader and John Abbotts, *The Menace of Atomic Energy,* rev. ed. (New York: Norton, 1979), p. 116.

16. Ibid., p. 115.

17. Ibid., p. 117–19.

18. *Win Magazine,* February 1, 1979.

19. Anna Gyorgy and Friends, *No Nukes* (Boston: South End Press, 1979) p. 338.

20. Berger, *The Unviable Option,* p. 57.

21. Richard E. Webb, *Accident Hazards of Nuclear Power Plants* (Amherst, Mass.: University of Massachusetts Press, 1976), p. 98.

22. Ibid.

23. Ibid., pp. 41–73.

24. Ibid., p. 41.

25. Berger, *The Unviable Option,* p. 60.

26. Ibid., p. 46.

27. Lindsay Mattison and Richard Daly, "A Quake at Bodega," *Nuclear Information,* April 1964, pp. 1–12.

28. *New York Times,* December 31, 1968.

29. *Life Magazine,* May 1979.

30. *New York Times,* January 31, 1979.

31. Berger, *The Unviable Option,* p. 144.

5. Nuclear Power: Myths and Realities

1. *New Age Magazine,* June 1979.

2. Anna Gyorgy and Friends, *No Nukes* (Boston: South End Press, 1979) p. 92.

3. Ralph Nader and John Abbotts, *The Menace of Atomic Energy,* rev. ed. (New York: Norton, 1979), p. 184.

4. Gyorgy, *No Nukes,* p. 85.

5. Peter Faulkner, *The Silent Bomb* (New York: Vintage, 1977), p. 239.

6. Ibid., pp. 23–40.

7. John J. Berger, *The Unviable Option* (New York: Dell, 1977), pp. 150–51.

8. *Electrical World,* September 15, 1977. In Gyorgy, *No Nukes,* p. 171.

9. Gyorgy, *No Nukes,* p. 192.

10. Gyorgy, *No Nukes,* p. 193.

11. Ibid.

12. Ibid., p. 191.

13. Berger, *The Unviable Option,* p. 138.

14. Ibid., p. 137.

15. Ibid., p. 139.

16. Gyorgy, *No Nukes,* p. 175.

17. Ibid., p. 173.

18. Berger, *The Unviable Option,* p. 144.

19. Nader and Abbotts, *Menace of Nuclear Power,* p. 228.

20. Ibid., p. 227.

21. Gyorgy, *No Nukes,* p. 180.

22. Ibid., p. 183.

23. Ibid.

24. Ibid., p. 179.

25. Harvey Wasserman, "Anti Nuclear Movement Approaches Critical Mass," *New Age Magazine,* June 1979.

26. Gyorgy, *No Nukes,* p. 216.

27. All testimony in this chapter, except where noted, is from the U.S. Congress, Joint Committee on Atomic Energy, *Atomic Energy Commission Authorizing Legislation, Fiscal Year 1968,* Part 2, March 14 and 15, 1967, pp. 741–68.

28. Ibid., p. 744.

29. Ibid., p. 748.

30. Ibid., p. 883, Appendix 13, pp. 1296, 1298.

31. U.S. Congress, Joint Committee on Atomic Energy, *Licensing and Regulation of Nuclear Reactors,* Part 1 (April 4, 5, 6, 29, and May 3, 1967) or Part 2 (September 12, 13, and 14, 1967), p. 18.

32. Ibid., p. 90.

33. U.S. Congress, Joint Committee on Atomic Energy, *Radiation Safety and Regulation,* June 15, 1961, p. 366.

6. Human Frailty and Inhuman Technology

1. U.S. Congress, Joint Committee on Atomic Energy, *Loss of the USS Thresher,* p. ix.

2. Ibid., p. 128.

3. Water-Reactor Safety Program Office, Phillips Petroleum Company, *Draft of Water-Reactor Safety Program: Summary Description.* In U.S. Congress, Joint Committee on Atomic Energy, *Atomic Energy Commission Authorizing Legislation, Fiscal Year 1968,* Part 2, March 14 and 15, 1967, p. 1369.

4. Donald Oken, "Mental Preparedness of Emergency Personnel." In Laurence Lanzl, John H. Pinge, and John H. Rust, eds., *Radiation Accidents and Emergencies* (Springfield, Ill.: Charles C. Thomas, 1965).

5. T. J. Thompson and J. G. Beckerley, eds., *The Technology of Nuclear Power,* vol. 1 (Cambridge, Mass.: M.I.T. Press, 1964), pp. 700–1.

6. Ibid., p. 7.

7. Ibid., p. 5.

8. W. B. Lewis, "Report CRR-836," and D. G. Hurst, "Report GPI 14." In Charles R. Russell, *Reactor Safeguards* (Elmsford, N.Y.: Pergamon Press, 1962).

9. Leo Goodman, "Radiation Hazard in Industry." Speech presented to the 32nd All-Ohio Safety Congress, Columbus, Ohio, April 17, 1962.

10. Case histories from *A Summary of Industrial Accidents in USAEC Facilities, 1961–1962,* TID-5360, Suppl. 4, December 1963.

11. Leo Goodman, "Radiation Hazard in Modern Industry." Speech presented to John Fogarty Memorial Luncheon, Washington, D.C., April 26, 1967.

12. Atomic Energy Commission, *Annual Report to Congress of the Atomic Energy Commission, 1966,* p. 35, and *1967,* p. 159.

13. Atomic Energy Commission, *Annual Report to Congress of the Atomic Energy Commission, 1967,* p. 284.

14. *New York Times,* August 30, 1966 and November 11, 1966.

15. Robert Gannon, "What Really Happened to the *Thresher,*" *Popular Science Monthly,* February 1964. See also Norman Polmar, *Death of the Thresher* (Radnor, Penn.: Chilton Books, 1964), pp. 116–17.

16. Atomic Energy Commission Division of Technical Information, *Nuclear Safety* (Washington: Atomic Energy Commission, January/February 1968), p. 89.

17. Siegel legal action taken from Brief for Petitioner and Joint Appendix in the United States Court of Appeals for the District of Columbia Circuit, no. 21, p. 342.

18. John J. Berger, *The Unviable Option* (New York: Dell, 1977), p. 205.

19. Ibid., p. 192.

20. Ibid., p. 193.

21. *Seven Days Magazine,* April 13, 1979.

22. Berger, *The Unviable Option,* p. 191.

23. Edward Teller, "How Shall Nuclear Technology be Applied?" In Mills, Biehl, and Mainhardt, eds., *Modern Nuclear Technology* (New York: McGraw-Hill, 1960), p. 306.

7. Near to the Madding Crowd

1. U.S. Congress, Joint Committee on Atomic Energy, *Licensing and Regulation of Nuclear Reactors,* Part 1 (April 4, 5, 6, 20, and May 3, 1967) or Part 2 (September 12, 13, and 14, 1967), p. 95. (Abbreviated from hereon as *L & R.*)

2. T. J. Thompson and J. G. Beckerley, eds., *The Technology of Nuclear Power,* vol. 1 (Cambridge, Mass.: M.I.T. Press, 1964), p. 6.

3. *National Coal Policy Conference Newsletter,* March 24, 1967.

4. Peter Faulkner, *The Silent Bomb* (New York: Vintage, 1977), p. 1.

5. Ibid., pp. 16–17.

6. Ibid., p. 11.

7. *New York Times,* November 6, 1967.

8. Clifford K. Beck, "Reactor Siting and Practice in the U.S." Speech presented at the American Nuclear Society, Los Angeles Section, February 16–18, 1965.

9. Ibid.

10. Thompson and Beckerley, *Technology of Nuclear Power*, p. 6.

11. Clifford K. Beck, "Engineering Out the Distance Factor." Speech before the 1963 Annual Convention of the Federal Bar Association.

12. Ibid.

13. *Daily Argus,* Mount Vernon, N.Y., July 18, 1963.

14. Ibid.

15. Consolidated Edison, statement of December 10, 1962.

16. Anna Gyorgy and Friends, *No Nukes* (Boston: South End Press, 1979), p. 206.

17. *Nucleonics,* May 1967. Quoted in Gyorgy, *No Nukes,* p. 206.

18. *L & R,* pp. 72–73.

19. Ibid., p. 70.

20. Ibid., p. 94.

8. No Place to Hide

1. *Boston Globe,* January 25, 1978. In Anna Gyorgy and Friends, *No Nukes* (Boston: South End Press, 1979), p. 43.

2. Ibid.

3. *New York Times,* May 23, 1964.

4. Ritchie Calder, *Living With The Atom* (Chicago: University of Chicago Press, 1962), pp. 89-90.

5. Charles R. Russell, *Reactor Safeguards* (Elmsford, N.Y.: Pergamon Press, 1962), pp. 275–76.

6. *New York Times,* May 16, 1968.

7. Gyorgy, *No Nukes,* pp. 363–65.

8. *New York Times,* March 14, 1965.

9. *Atlantic City Press,* May 16, 1968.

10. Gyorgy, *No Nukes,* p. 82 (for example).

11. Ibid., p. 42.

12. *New York Times,* February 23, 1965.

13. *New York Times,* January 25, 1967.

14. *New York Times,* November 18, 1967.

15. *New York Times,* November 25, 1967.

16. Atomic Energy Commission, *Annual Report for 1967* (Washington, D.C.: Atomic Energy Commission, 1967), p. 274.

17. Atomic Energy Commission, *Annual Report for 1966* (Washington, D.C.: Atomic Energy Commission, 1966), p. 274.

18. Atomic Energy Commission, *Annual Report for 1967* (Washington, D.C.: Atomic Energy Commission, 1967), p. 211.

19. Atomic Energy Commission, Division of Industrial Participation, *The Nuclear Industry 1966* (Washington, D.C.: Atomic Energy Commission), p. 161.

20. Gyorgy, *No Nukes*, p. 54.

21. Ibid.

22. Atomic Energy Commission, Release no. S-8-68, February 27, 1968.

23. Calder, *Living With The Atom,* p. 230.

24. Atomic Energy Commission, *The Nuclear Industry 1966.*

25. Ibid., p. 162.

26. Atomic Energy Commission, *A Summary of Industrial Accidents in U.S. Atomic Energy Commission Facilities 1963-1964,* TID-5360, Suppl. 5, December 1965, p. 15ff.

27. Ibid., pp. 1-7.

28. *New York Times,* April 11, 1968.

29. Gyorgy, *No Nukes*, p. 124.

30. Leo Goodman, "Radiation Hazard in Modern Industry." Speech presented to John Fogarty Memorial Luncheon, Washington, D.C., April 26, 1967.

31. *New York Times,* August 19, 1966 and August 20, 1966.

32. *New York Times,* September 9, 1966.

33. Goodman, "Radiation Hazard."

9. We Interrupt This Broadcast . . .

1. Detroit-Edison and Associates, "Information Report to the Project Companies of Dow Chemical," Nuclear Power Development Project TID-10077, December 1, 1953, p. 158.

2. Associated Press, May 11, 1979.

3. Associated Press, May 1979.

4. Ron Lanoue, "Evacuation Plans—The Achilles Heel of the Nuclear Industry," *Public Citizen*, 1975, p. 11.

5. *Atlantic City Press,* January 11, 1977.

6. Laurence Lanzl, John H. Pinge, and John H. Rust, eds., *Radiation Accidents and Emergencies* (Springfield, Ill.: Charles C. Thomas, 1965), p. 126.

7. C. R. McCullough, ed., *Safety Aspects of Nuclear Reactors* (New York: Van Nostrand, 1957), p. 157.

8. Tom Stonier, *Nuclear Disaster* (Cleveland: World Publishing, 1964), pp. 75-76.

10. The Thousand-Year Curse

1. Jack Schubert and Ralph E. Lapp, *Radiation* (New York: Viking Press, 1957), p. 269.

2. Ritchie Calder, *Living With The Atom* (Chicago: University of Chicago Press, 1962), pp. 89-90.

3. *Nucleonics,* December 1966, p. 48.

4. Anna Gyorgy and Friends, *No Nukes* (Boston: South End Press, 1979), pp. 50–52.

5. Buttermilk Creek data from Rochester Committee for Scientific Information, Bulletin No. 2, February 24, 1968 and Bulletin No. 3, February 28, 1968.

6. *Nucleonics,* December 1966.

7. Gyorgy, *No Nukes,* p. 52.

8. Walter Schneir, "The Atom's Poisonous Garbage," *The Reporter,* March 17, 1960.

9. United Mine Workers of America, "The Atom . . . Friend or Foe?" *UMW Journal,* p. 21.

10. McKinley C. Olson, *Unacceptable Risk,* (New York: Bantam, 1976), p. 91.

11. Ibid.

12. Ibid., p. 89.

13. Ibid., p. 92.

14. Gyorgy, *No Nukes,* p. 128.

15. Ibid., p. 126.

16. John J. Berger, *The Unviable Option* (New York: Dell, 1977), pp. 102-3.

17. Gyorgy, *No Nukes,* p. 127.

18. David E. Lilienthal, "When the Atom Moves Next Door," *McCall's,* October 1963.

19. U.S. Congress, Joint Committee on Atomic Energy, *Atomic Energy Authorizing Legislation, Fiscal Year 1968,* Part 2, March 14, and 15, *1967,* p. 935.

20. Calder, *Living With The Atom,* pp. 203-4.

21. Wilfred E. Johnson. Speech before the Health Physics Society, Augusta, Georgia, January 24, 1968.

22. Joel A. Snow, "Radioactive Waste from Reactors," *Scientist and Citizen Magazine,* May 1967.

23. United Mine Workers, "The Atom . . . Friend or Foe?" p. 25.

24. Rachel Carson, *Silent Spring* (Boston: Houghton Mifflin, 1962), p. 168-69.

11. Thermal Pollution and Man's Dwindling Radiation Budget

1. Robert H. Boyle, "A Stink of Dead Stripers," *Sports Illustrated,* April 26, 1965, p. 81.

2. W. Donham Crawford. Speech at White Plains Rally, October 19, 1968.

3. U.S. Congress, House Committee on Science and Astronautics, *Report on the International Biological Program* (Washington, D.C., March 1968).

4. Gerald F. Tape. Speech presented at the annual meeting of the Southern Interstate Nuclear Board, Hot Springs, Arkansas, April 1, 1968. AEC Release No. S 12-68, April 1, 1968.

5. Anna Gyorgy and Friends, *No Nukes* (Boston: South End Press, 1979), pp. 385–86.

6. Boyle, "A Stink of Dead Stripers."

7. U.S. Congress, Joint Committee on Atomic Energy, *Licensing and Regulation of Nuclear Reactors,* Part 1, April 4, 5, 6, 20, and May 3, 1967, or Part 2, September 12, 13, and 14, 1967, p. 27.

8. Richard Curtis and Elizabeth Hogan, *Perils of the Peaceful Atom,* 1st ed. (New York: Doubleday, 1969), pp. 146–47.

9. Barry Commoner, "Nuclear Power and Environment: An Inquiry." Speech before a conference on nuclear power and environment, Stratton Mountain, Vermont, September 11 and 12, 1968.

10. Malcolm Peterson, "Krypton-85, Nuclear Air Pollutant," *Scientist and Citizen Magazine,* March 1967.

11. Ralph Nader and John Abbotts, *The Menace of Atomic Energy,* rev. ed. (New York: Norton, 1979), p. 73.

12. Norman Landsdell, *The Atom and the Energy Revolution* (New York: Philosophical Library, 1958), p. 173.

13. Discussion of zinc 65 from Malcolm Peterson, "Environmental Contamination from Nuclear Reactors," *Scientist and Citizen Magazine,* November 1965.

14. Peter Faulkner, *The Silent Bomb* (New York: Vintage, 1977), p. 124.

15. Gyorgy, *No Nukes,* p. 56.

16. Ritchie Calder, *Living With The Atom* (Chicago: University of Chicago Press, 1962), p. 268.

17. Walter Schneir, "The Atom's Poisonous Garbage," *The Reporter,* March 17, 1960.

18. Conversation reported by Ann Carl, March 25, 1968 in her report on the Health Physics Society Symposium, January 24–26, 1968, in Augusta, Georgia.

19. Dorothy Nelkin, *Nuclear Power and Its Critics* (Ithaca, N.Y.: Cornell University Press, 1971).

20. John J. Berger, *The Unviable Option* (New York: Dell, 1977), p. 68.

21. Ibid., p. 67.

22. Anthony Netboy, "Nuclear Power on Salmon Rivers," *The Nation,* October 9, 1967.

23. Lamont C. Cole, "Can The World Be Saved?" *New York Times Magazine,* March 31, 1968.

12. Absolute Power

1. Ralph Nader and John Abbotts, *The Menace of Atomic Energy* (New York: Norton, 1979), p. 278.

2. Ibid., p. 279.

3. Ibid.

4. Ibid., p. 277.

5. Ibid.

6. Peter Faulkner, *The Silent Bomb* (New York: Vintage, 1977), p. 281.

7. Anna Gyorgy and Friends, *No Nukes* (Boston: South End Press, 1979), p. 110.

8. Faulkner, *Silent Bomb,* pp. 315–16.

9. Ibid., pp. 184–85.

10. Harold P. Green, "'Reasonable Assurance' of 'No Undue Risk,'" *Notre Dame Lawyer,* June 1968. Reprinted in *Scientist and Citizen Magazine,* June/July 1968.

11. Ibid.

12. Saul Friedman, "The Enrico Fermi Power Plant," *Detroit Free Press,* July 17, 1966.

13. T. J. Thompson and J. G. Beckerley, eds., *The Technology of Nuclear Reactor Safety,* vol. 1 (Cambridge, Mass.: M.I.T. Press, 1964), p. 4.

14. U.S. Congress, Joint Committee on Atomic Energy, *Licensing and Regulation of Nuclear Reactors,* Part 1, April 4, 5, 6, 20, and May 3, 1967, or Part 2, September 12, 13, and 14, 1967, p. 81. (Abbreviated from hereon as L & R.)

15. Atomic Energy Commission, Release No. L-109, May 28, 1968.

16. Faulkner, *Silent Bomb,* p. 237.

17. *Life Magazine,* May 1979.

18. *L & R,* p. 36.

19. Ibid., p. 7–8.

20. Ibid., pp. 72–73.

21. Ibid., p. 129.

22. Ibid., p. 113.

23. Ibid., p. 37.

24. Ibid., p. 133.

25. T. E. Murray, *War and Peace* (Cleveland: World Publishing, 1960), pp. 200–2.

26. Assembly of the State of California, Assembly Interim Committee on Industrial Relations, "Nuclear Safety." Published as part of *Assembly Interim Committee Reports 1965–1967* (1968).

27. *L & R,* p. 27.

28. *Congressional Record,* July 16, 1963.

29. *National Coal Policy Conference Newsletter,* December 16, 1965. Also, Nos. 22603-22607 of Commerce Clearing House, Atomic Energy Law Reports, November 15, 1965, order of New Jersey Board of Public Utility Commissioners, and Summary of April 22, 1966, Interim Order.

30. Gyorgy, *No Nukes,* pp. 21–22.

31. *L & R,* p. 37.

32. Green, "Reasonable Assurance."

33. *L & R,* pp. 30–31.

34. Atomic Energy Commission, "In The Matter of Consolidated Edison Company of New York, Inc." Petition to Intervene by the Conservation Center, Inc., Docket No. 50-247.

35. W. B. McCool, Atomic Energy Commission, Memorandum and Order, AEC Docket No. 50-247, issued on December 20, 1966.

13. The Scandal of Insurance and Subsidies

1. U.S. Congress, Joint Committee on Atomic Energy, *Hearings on Governmental Indemnity*, 84th Congress, 2d session, pp. 248–50.

2. "Nuclear Insurers Confident," *New York Times*, March 30, 1979.

3. Ralph Nader and John Abbotts, *The Menace of Atomic Energy*, rev. ed. (New York: Norton, 1979), p. 99.

4. Peter Faulkner, *The Silent Bomb* (New York: Vintage, 1977), p. 160.

5. Ibid., p. 161.

6. Ibid.

7. Nader and Abbotts, *Menace of Atomic Energy*, p. 227.

8. Richard Curtis and Elizabeth Hogan, *Perils of the Peaceful Atom* (New York: Doubleday, 1969), p. 196.

9. *Congressional Record,* August 4, 1965.

10. Curtis and Hogan, *Perils,* p. 199.

11. Nader and Abbotts, *Menace of Atomic Energy,* pp. 28, 370.

12. J. F. terHorst, "Are We Willing to Pay for Nuclear Accidents?" *St. Petersburg Times,* April 25, 1979.

13. WBAI News (New York City), July 2, 1979.

14. terHorst, "Are We Willing to Pay?"

15. Alan Berlow, "Atomic Accident Liability," *St. Petersburg Times,* April 19, 1979.

16. Mike Gravel, "Finding the Critical Mass."

17. U.S. Congress, Joint Committee on Atomic Energy, *Hearings on Governmental Indemnity,* p. 251.

18. Ibid., p. 252.

19. Jack Anderson, syndicated column, May 11, 1979.

20. Ibid., November 1977.

21. *St. Petersburg Times,* August 5, 1979.

22. Anna Gyorgy and Friends, *No Nukes* (Boston: South End Press, 1979), p. 179.

23. Ibid., p. 180.

24. Ibid., p. 173.

25. *Life Magazine,* May 1979.

26. Curtis and Hogan, *Perils,* pp. 227–28.

27. Gyorgy, *No Nukes,* p. 183.

28. Curtis and Hogan, *Perils,* p. 182.

29. Gyorgy, *No Nukes,* p. 175.

30. Shad Alliance, "Shoreham: $2 billion of Dangerous Obsolescence," 1978.

31. John J. Berger, *The Unviable Option* (New York: Dell, 1977), p. 137.

32. Gyorgy, *No Nukes,* p. 176.

33. Ibid., p. 48.

14. From Fissile to Fizzle

1. Richard Curtis and Elizabeth Hogan, *Perils of the Peaceful Atom* (New York, Doubleday, 1969), p. 201.

2. *Nucleonics Week* and *NuExco,* personal communication, July 1979.

3. John J. Berger, *The Unviable Option* (New York: Dell, 1977), p. 116.

4. Anna Gyorgy and Friends, *No Nukes,* (Boston: South End Press, 1979), p. 372.

5. Ralph Nader and John Abbotts, *The Menace of Atomic Energy,* rev. ed. (New York: Norton, 1979), p. 226.

6. Berger, *Unviable Option,* p. 116.

7. Gyorgy, *No Nukes,* p. 172.

8. Ibid., p. 151.

9. Ibid.

10. Nader and Abbotts, *Menace of Atomic Energy,* p. 89.

11. Berger, *Unviable Option,* p. 115.

12. Ibid.

13. Ibid., pp. 117–28.

14. Curtis and Hogan, *Perils,* p. 209.

16. Citizen Action

1. *New York Times,* April 28, 1970.

2. *New York Times,* May 7, 1967.

3. Robert Rienow, *American Problems Today,* (Lexington, Mass.: D.C. Heath, 1953).

4. *New York Times,* April 4, 1971.

5. Leonard Ross, "How 'Atoms for Peace' Became Bombs for Sale," *New York Times Magazine,* December 5, 1976.

Index

About the Authors

Richard Curtis was born in New York City in 1937. He received degrees from Syracuse University and the University of Wyoming. He has written over forty works of nonfiction and fiction, including a satire on environmental issues called *The Case for Extinction*. He lives in New York City where he is the president of a successful literary agency.

Elizabeth Hogan was born and educated in Philadelphia. She is the author of *The Power of Words in Your Life*. A member of Common Cause, Friends of the Earth, Natural Resources Defense Council, and Public Citizen, she currently lives in St. Petersburg, Florida.

Shel Horowitz was born in New York City in 1956. A graduate of Antioch College, he is a journalist, community activist, and poet. He has been involved with the antinuclear power and weapons movement since 1972, and he has written on the subject for such publications as the *Providence Evening Bulletin, East Side-West Side, Win Magazine,* and *WBAI Folio*.

Richard E. Webb holds a Ph.D. in Nuclear Engineering from Ohio State University. Serving four years in the Division of Naval Reactors Commission as a junior reactor engineer, he was responsible for the nuclear reactor portion of the Shippingport pressurized water reactor. He has been employed by both the Atomic Energy Commission and by private industry in reactor research and development. For the past ten years, Dr. Webb has been raising reactor safety issues in federal and Congressional hearings. He is the author of *The Accident Hazards of Nuclear Power Plants*, University of Massachusetts Press, 1976.